SATELLITE-BASED GLOBAL CELLULAR COMMUNICATIONS

Other McGraw-Hill Telecommunications Books of Interest

Ali   *Digital Switching Systems*
Bartlett   *Cable Communications*
Benner   *Fibre Channel*
Best   *Phase-locked Loops, 3d Edition*
Goldberg   *Digital Techniques in Frequency Synthesis*
Goralski   *Introduction to ATM Networks*
Harte   *Cellular + PCS: The Big Picture*
Harte   *CDMA IS-95*
Heldman   *Computer Telecommunications*
Heldman   *Information Telecommunications Millenium*
Inglis   *Video Engineering, Second Edition*
Kessler   *ISDN, Second Edition*
Lee   *Mobile Cellular Telecommunications, Second Edition*
Lee   *Mobile Communications Engineering, Second Edition*
Levine   *GSM Superphones*
Lindberg   *Digital Broadband Networks and Services*
Logsdon   *Mobile Communication Satellites*
Macario   *Cellular Radio, Second Edition*
Pecar   *Telecommunications Factbook*
Roddy   *Satellite Communications, Second Edition*
Rohde et al.   *Communication Receivers, Second Edition*
Simon et al.   *Spread Spectrum Communications Handbook*
Smith & Gervelis   *Cellular System Design and Optimization*
Tsakalakis   *PCS Network Deployment*
Winch   *Telecommunications Transmission Systems*

To order or receive additional information on these or any other McGraw-Hill titles in the United States, please call 1-800-722-4726. Or visit us at www.ee.mcgraw-hill.com. In other countries, contact your local McGraw-Hill representative.

Satellite-Based Global Cellular Communications

Bruno Pattan

Senior Member of the Technical Staff
Federal Communications Commission
Office of Engineering and Technology

McGraw-Hill
New York • San Francisco • Washington, D.C. • Auckland • Bogotá
Caracas • Lisbon • London • Madrid • Mexico City • Milan
Montreal • New Delhi • San Juan • Singapore
Sydney • Tokyo • Toronto

Library of Congress Cataloging-in-Publication Data

Pattan, Bruno.
    Satellite-based global cellular communications / Bruno Pattan.
        p.   cm.
    Includes bibliographical references and index.
    ISBN 0-07-049417-7
    1. Artificial satellites in telecommunication.   2. Global system
for mobile communications.   3. Cellular radio.   I. Title.
    TK5104.P375   1998
    621.3845′6—dc21   97-26254
                                                                           CIP

McGraw-Hill

A Division of The **McGraw·Hill** Companies

1 2 3 4 5 6 7 8 9 0 FGR/FGR 9 0 2 1 0 9 8 7

ISBN 0-07-049417-7

*The sponsoring editor for this book was Stephen S. Chapman, the editing supervisor was
Sharon Neal, and the production supervisor was Sherri Souffrance. It was set in Vendome
by North Market Street Graphics.*

Printed and bound by Quebecor/Fairfield

McGraw-Hill books are available at special quantity discounts to use as
premiums and sales promotions, or for use in corporate training programs. For
more information, please write to the Director of Special Sales, McGraw-Hill,
11 West 19th Street, New York, NY 10011. Or contact your local bookstore.

 This book is printed on recycled, acid-free paper containing a minimum
of 50% recycled, de-inked fiber.

The views or conclusions contained in this book are those of the author and should not be intepreted as necessarily representing the official policies, either expressed or implied, of the Federal Communications Commission.

CONTENTS

Contents

PREFACE

The explosive growth in terrestrial wireless communications, both indigenously and internationally, has evoked an economic interest in providing an extension of this growth to satellite-based mobile communications. Its genesis was approximately twenty years ago with the advent of satellites in geostationary orbit (h = 35,000 km) providing maritime services. The ground terminals are dedicated to a particular service, are expensive, and do not provide personal communications services.

As the satellite system technology developed, it became possible to provide personal service by the use of satellites in orbital regimes other than geostationary orbit. This included low earth orbit (LEOs at about 1000 km altitude), medium earth orbit (MEOs at about 10,000 km altitude), and even the highly elliptical orbit of the Molniya-type with apogee in the northern hemisphere.

This book addresses this relatively new mode of personal communications via satellite-based cellular communications. The concepts are not unlike that employed by terrestrial cellular systems, but there are differences which will be discussed in this book.

Topics include mechanics of orbital flight in the low orbit regime, cellular structuring, propagation for wireless, information on the ground terminals which consists of handheld transceivers, and coverage requirements for satellites which are visible for short periods of time and located anywhere in the sky. However, the all-important frequency bands to transmit and receive (without which there would be no service) have not been neglected.

This book may be considered technical in some circles, but many sections are of a general nature. In totality though, it is intended for system engineers, analysts, program managers, technicians, and other technical personnel with a need to understand satellite-based wireless communications. To the thousands in the world who are now working in terrestrial systems, both wireline and cellular, this book is essential. Before the next millennium arrives, this field is where the market and competition is coming from.

Bruno Pattan

CHAPTER **1**

Introduction

In the coming years, prior to the next millennium, a number of consortia are prepared to launch constellations of satellites which will provide continuous and global data and telephony service. This impetus came about due to the burgeoning market for and development of terrestrial cellular systems and the availability of new frequency bands for this service. The satellite antennas can project patterns of cells onto the Earth's surface, thereby emulating their terrestrial counterparts. The satellites will be principally in low earth orbit. The benefits envisioned are manifold. These include moderate launch costs because of the low-orbit regime, lower time delays avoiding the (at least) one-half second delay, the double talk (echoes) problems which afflict satellites in GSO (altitude of 35,000 km), less complicated satellite configurations, and lower radiated power (by as much as 30 dB). Also, from a market point of view, the satellites will provide service where cellular systems do not exist and probably never will.

On the ground segment side, communication will be possible with personal handsets providing data and full duplex voice service, even though the handsets are inherently low-performance devices because of their low-antenna gain and limited power output.

The advantages of satellite service using low-orbit regimes certainly does not mean that the GSOs are dinosaurs. There will always be a niche for both LEOs and GSOs depending on their applications.

On the downside, an ensemble of satellites is required in orbit to provide continuous and ubiquitous service because any single satellite will only remain in view for a small part of its orbit time. In addition, the earth coverage of a single satellite is only about 5 to 10 percent so that many are required for full global coverage. The implication of this last factor is that the global/continuous service is only possible if the full constellation is in orbit with its multiple-control earth stations; and phased rollout is not possible. In addition to the space segment, the ground networking plays a vital role and is more daunting than for satellites in GSO which are basically stationary. Finally, but certainly no less important, is the cost to put these systems into service. The cost may vary from several hundred million to several billion dollars, depending on the service to be provided. Generally, satellites providing voice service are more complicated than only data service systems.

Therefore, the low earth orbit satellites are coming to provide telecommunications service and will complement, not supplant, the terrestrial cellular systems and/or the existing wire line telephone infrastructure.

Terrestrial cellular service and wire line service are restricted to geographical areas, as opposed to satellite-based service where potential customers may be located anywhere (a truly ubiquitous service).

In regions where population density is low or where economics dictate, wire line (much less cellular) service is not economically viable; space-based wireless services will arrive faster and at much less cost. These gaps may be filled in and may even become the main source of communication in the region. In particular, this may include many parts of Africa, Asia, and South America.

Wireless communication is undergoing an explosive growth period, and satellite-based delivery will become a major player in this near-revolutionary change.

The following chapters will present many of the peculiarities and subtleties of this dynamic area of satellite-based wireless cellular communication.

Nongeostationary Orbit (Non-GSO) Satellites

2.1 Introduction to Nongeostationary Satellites

Satellites using low earth orbits are finding increased uses for new services in recent years. These nongeostationary orbits (NGSO)* may have circular orbits up to 10,000 km altitude. These services include surveillance, earth resource observations, scientific microgravity experiments, and, more recently, telecommunications. Previously, the most dominant domain of communications satellites has been in geostationary orbit. Our early venture into communication satellites used satellites in low earth orbit. Clearly there were reasons for this mode of operation. First, the booster technology was not available to loft satellites into a higher orbit. Second, the satellite technology was not sufficiently mature to realize higher performance such as power generation and signal bandwidth. Also, there was a problem with reliability since operating in a hostile space environment was mostly unknown. These satellites were short-lived and getting them into orbit in itself was an accomplishment.

It is the concurrence of many (but not all) that indeed non-GSOs may have their telecommunications niche and can perform valuable functions which can complement the high performers in geostationary orbit. At the present time, some of these niches may be solely the domain of non-GSOs. This would include voice and data services to or from handheld transceivers and the provision of global cellular coverage.

However, a renaissance in low earth communication satellites has occurred, and they are used to provide voice and data services domestically and globally. The services involved are relatively new and include cellular telephones on a global basis, position determination, search and rescue, data, and electronic mail (store and forward) for worldwide transfer.

This new breed of satellite has been referred to as lightsats, microsats, and even cheapsats. A strict definition of a telecommunication lightsat has not been established. In some circles, it has been defined as a satellite weighing less than 500 kg (1100 lb), the size of a desk (or smaller), and lofted into an orbit between 700 km and 10,000 km in height. The lower altitude is dictated by the atmospheric drag and location at the higher altitude by the Van Allen radiation belts (two) which are detrimental to the onboard electronics. Cost is an additional factor in the definition of a

---

* The ITU (International Telecommunications Union) has designated satellites which are not in geostationary orbit as non-GSO satellites. This includes low earth orbit (LEO) satellites (up to about 2000 km altitude), medium earth orbit (MEO) satellites (up to about 20,000 km altitude), and highly elliptical orbits (HEOs) with apogees up to 40,000 km and beyond.

low-orbit satellite and is in the range of \$1 to \$10 million. However, for satellites with substantial performance, this figure is on the low side. In particular, those satellites provided data and voice service.

A broader definition of lightsat should include a satellite at higher altitudes, even up to geostationary orbit. Several U.S. companies are considering lightsats for this altitude but with reduced capacity from the leviathans which are already up there. These would appear to be more expensive (both the satellite and launching it) than the ones operating in low altitude.

Generally, all non-GSOs have capabilities which are far less than the communication satellites operating in geostationary orbit. For example, GSOs have 24 transponders or more where satellites in non-GSO may support possibly a dozen or so. Another factor which limits the capacity of non-GSOs is the paucity of spectra available.* In the United States, several organizations which also include the terrestrial users are vying for use of the same spectrum. Interference is a problem and often can only be solved by band segmentation or dividing up the spectrum to be used by a single user on an exclusive basis, with at least the assurance of noninterference of operation if others are allocated the same spectrum.

Several technological advances have also accelerated this non-GSO satellite renaissance. This includes the miniaturization of onboard electronics, RF components, and computers making possible high-density circuits with high performance. In addition, the ground earth station transceivers have been reduced in size to the point where they are handheld. This technology has been accelerated by the advent of terrestrial cellular systems. On a more macroscopic scale, the maturation of low-cost small launch vehicles capable of launching the lightsats into low earth orbit are now available.

There are several advantages and disadvantages accrued by the use of low earth orbit satellites for telecommunications application. These are listed in Tables 2.1 and 2.2.

2.1.1 Communication Modes of Operation

As indicated previously, for single satellite operation two modes of communication between two sites are possible. One is the bent-pipe relay

---

* LEO lightsats providing communication service operate mostly in the VHF and UHF bands and lower microwave band. In the United States, the former uses of the band are designated as small LEOs and provide data but no voice service. The lower microwave band for communications is used by the big LEOs which provide both data and voice service.

TABLE 2.1

Advantages of
Non-GSO Satellites

- Less booster power is required to put a non-GSO satellite into orbit (multiple rocket stages not required) than that used for a GSO satellite.

- Operating at lower altitudes (<10,000 km) reduces the space loss for signal propagation. This attribute lowers the onboard power requirements; thus, fewer solar cells and power packs are required. The ground facility design is also simplified. For example, for the type of propagation losses incurred as a function of distance and frequency, see Fig. 2.1.

- Simplified satellite altitude control is possible since passive gravity gradient stabilization can be used with possible passive or active magnetic devices interacting with earth's magnetic field. Gravity gradient stabilization is not feasible at GSO altitudes.

- For speech communications, the problem of echoes (a speaker hearing his voice returned by the satellite) is not a significant problem because of the smaller delay than those found at GSO distances (e.g., GSO > 270 ms, NGSO < 100 ms).

- Cost is much lower to build and launch an NGSO satellite.

- Service can be provided to the higher latitudes (+81° the limit of GSOs) using polar orbits or high-inclination orbits.

- Their smaller size and weight permits multiple spacecraft to be launched by a single launch vehicle.

- VHF and UHF frequency of operation of some NGSOs permits use of low-cost wire antennas.

- Service can be provided to handheld transceivers.

- Availability of Doppler shift provides position determination.

mode[*] in which both terrestrial facilities are within line of sight of the low-orbit satellite. This is indicated in Fig. 2.2a. The mutual communication visibility is a function of the satellite altitude and minimum required elevation angle α of the ground facility's antenna beam axis. The elevation angle normally is dictated by the acceptable level of the multipath signals. This angle may be in the range of 10 to 20°. It is also noted that sharing the same window requires that both sites lie in the satellite antenna footprint. The maximum time and length of visibility is realized when the earth station lies on the satellite ground track.

The second mode is the store-and-forward communication as shown in Fig. 2.2b. There is no mutual visibility of the satellite by either earth station at site A or site B. When the satellite is in view of site A, the message is transmitted to the satellite and put into memory in its onboard computer. At some later programmed time, which could be at the opposite

---

[*] Real-time duplex communications between earth stations.

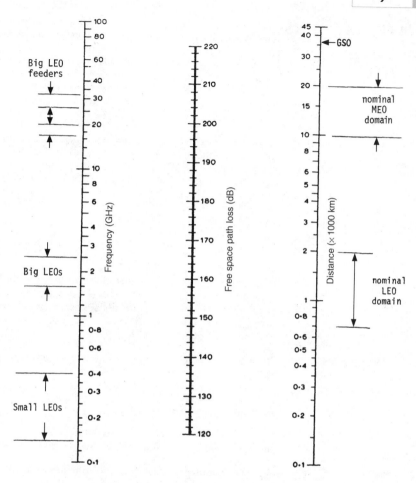

Figure 2.1
Nomogram relating frequency, free space loss, and range.

- Note the additional loss incurred in going from LEO ($h \approx 1000$ km) operation to GSO is about 30 dB.

side of the world, the message is retrieved by site B. The information is dumped within communication view of site B.

For an earth station which is coplanar (in ground track) with the satellite, the angular length of visibility for point A is $2\theta/360°$, where θ is one-half the earth central angle (see Fig. 2.2). The length of visibility is therefore $(2\theta/360°)T$, where T is the orbital period. It is noted as the communication visibility angle α ($\alpha = 0°$ for LOS) becomes larger, the central angle 2θ becomes smaller and the communication visibility becomes smaller. Clearly, when $\alpha = 0°$, the optical visibility is maximum, but the communication visibility is degraded at these low angles. For example, for a nominal altitude of 1000 km and $\alpha = 0°$, the maximum visibility time is

TABLE 2.2

Disadvantages of
Non-GSOs

- One hundred percent real-time communications cannot be provided by a single non-GSO satellite since it goes out of sight for a large percentage of its orbit. Many satellites are required in orbit to maintain continuous bent-pipe relay service. A single satellite can be used as a bent-pipe relay mode if both ground segments are within line of sight during the transmission period. However, this time is small, and length depends on the altitude of the satellite. Otherwise, messages can be delivered on a store-and-forward basis when there is no mutual visibility.

- Because of their rapid movement relative to the ground transceiver, the signals provided by NGSOs will undergo Doppler effects which require compensation. The earth's rotation also makes a contribution to the Doppler effect. The amount of correction is a function of the satellite altitude, frequency used, and satellite flight path with respect to the transmitted signal (uplink and downlink).

- Since the satellite is generally eclipsed several times per day by the earth, an onboard battery pack must be robust enough to carry the load during the periods without sun. The eclipse time can be reduced by raising the satellite altitude, if the mission requirements permit. For some systems, power is turned off during this period for conservation.

- The earth station antenna must track the rapidly moving satellite in order to be able to communicate. This problem can be alleviated by using antennas with near hemispherical coverage,* or using directional steerable (mechanical or electronic) antenna beams. The directional antennas may require handover to another satellite if continuous service is required.

- There is more complex network management because there is a constellation of satellites providing continuous and global coverage.

- Each satellite sees a small percentage of the earth and multiple satellites are required to obtain greater coverage.

- Useful lifetimes may be 5 to 10 years. But this has yet to be proven since they have no heritage. A satellite's lifetime in GSO lifetime is now 10 to 15 years.

- Orbital debris may become a real problem. As global systems (constellations) proliferate, hundreds will be put into low earth orbit and, as the satellites become inoperative, may become a problem.

---

*Hemispherical and omnidirectional coverage are not quite tantamount since generally omnis have a null on-axis (bifolium pattern). However, in practical antennas, there is some fill-in in the dip giving almost hemispherical coverage.

about 18 minutes for a period of 105 minutes. For $\alpha = 10°$ (a practical elevation angle of the earth station beam axis), the visibility is reduced to about 13 minutes. See, for example, Fig. 2.3.

2.1.2 Single-Plane Satellite Cluster Operation

For operational situations in which almost continuous real-time global coverage is required, several satellites in different orbits are required, while

Figure 2.2

Two transmission modes for satellites in low earth orbit.

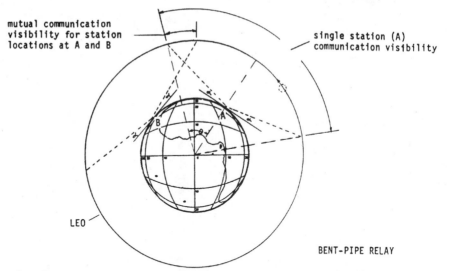

mutual communication visibility for station locations at A and B

single station (A) communication visibility

LEO

BENT-PIPE RELAY

(a) The satellite can be used in a "bent pipe" relay mode if both earth stations are within communication line-of-sight during the transmission/ receive period. The length of this period is a function of satellite altitude, beamwidth of satellite antenna (Rec/Xmit), and usable elevation angle and beam pointing/beamwidth of earth station antennas.

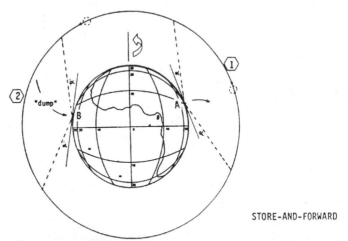

"dump"

STORE-AND-FORWARD

(b) The satellite is not mutually communication visible to stations A and B. Here, "store-and-forward" communications is used. Information uplinked to satellite in arc ① from station A and forwarded (dumped) to station B in arc ② . Satellite beams must be sufficiently broad to accept information in arc ① and also broad to dump anywhere in arc ② . Same applies to earth stations , unless a directional steerable beam is used.

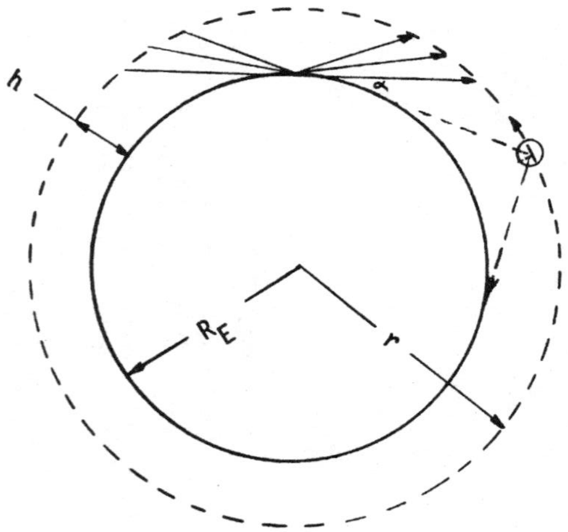

Altitude (km)	Elevation Angle, α		
	Time in View (minutes)		
	0°	10°	20°
800	15.3	11	7.5
1000	17.5	12.6	9.1
1200	20	15	11
1800	26.5	20.1	15.8
10,000	130	111	98.5

α = elevation angle of ground sensor beam axis

R_E = earth's radius

GM = gravitational parameter

G = universal gravitational constant

M = earth's mass

$$T = \sqrt{\frac{4\pi^2 r^3}{u}} \quad \text{period of circular orbit}$$

$$= 2\pi \sqrt{\frac{(R_E + h)^3}{GM}}$$

Note: $u = GM$.

1 km = 0.62 mi

Figure 2.3 Line-of-sight viewing times for satellites in LEO.

for uninterrupted *swath* coverage several satellites in the same orbit are required. A swath operation is depicted in Fig. 2.4. Clearly, clustering is required because satellites go out of sight for a large percentage of their flight.

For a given circular antenna footprint directed at the satellite subpoint (nadir), the maximum area covered is a function of the antenna beam width and satellite altitude. The percentage of the earth which can be viewed from a satellite as a function of altitude is indicated in Fig. 2.5. The earth station antenna elevation angle is parametric.

Figure 2.4

Depiction of a cluster of LEO satellites in a single orbit in which continuity of service in a given swath can be maintained. The swath features change due to the rotation of the earth (15° per hour).

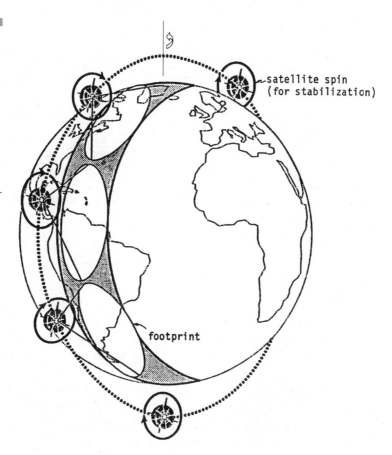

satellite spin
(for stabilization)

footprint

- No nodal regression for i=90°.

- The satellites are phased to maintain their position in orbit by the use of station keeping thrusters (e.g., using N₂ cold gas propulsion).

Figure 2.5

Percentage of earth coverage as a function of satellite orbital altitude, for various earth station antenna beam elevation angles.

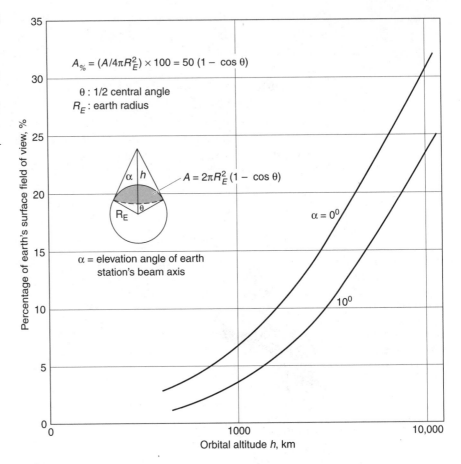

Figure 2.6 shows the visual swath and FOV angle as seen from a satellite as a function of the satellite's altitude. It is noted that for low earth orbit altitudes, the field-of-view angles are quite large, and the satellite antenna beam must be broad (with a sacrifice in gain) if full coverage of the swath is required. Generally, however, communication swaths are narrower and provide higher antenna gain to a swath with higher elevation angles at the periphery of the coverage area. If one extrapolates curve A to geostationary altitude (35,600 km), the FOV angle becomes 17.3°, visual swath central angle $2\theta = 2 \times 81.3 = 162.6°$, and swath width is 18,100 km. This is the great circle swath width and clearly not one-half the earth's circumference ($C = 40,074$ km) since the GSO satellite is located at a finite distance from the earth and not at infinity.

Orbits which are used in NGSO altitudes include inclined circular orbits (mostly with inclinations between 0° and 90°), circular polar orbits

Figure 2.6

Field of view and visual swath as a function of satellite altitude.

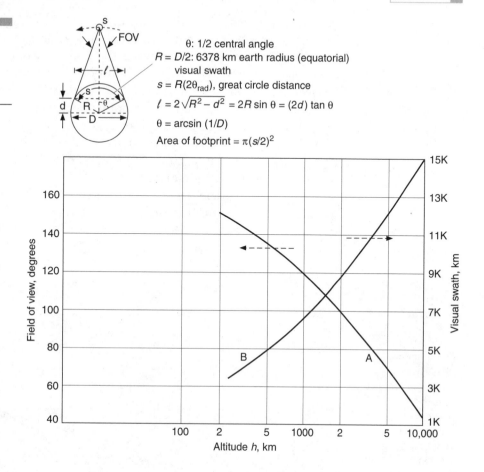

θ: 1/2 central angle
$R = D/2$: 6378 km earth radius (equatorial)
visual swath
$s = R(2\theta_{rad})$, great circle distance
$l = 2\sqrt{R^2 - d^2} = 2R\sin\theta = (2d)\tan\theta$
$\theta = \arcsin(1/D)$
Area of footprint $= \pi(s/2)^2$

with inclinations of 90° or near 90°, and elliptical orbits mostly to supply service to areas at the higher latitudes. The inclined orbits generally can also supply coverage at the high latitudes, and the maximum latitudinal excursion of their ground tracks is equal to the inclination angle. However, this does not mean that this is the farthest north that it can provide service. The antenna footprints can illuminate regions beyond this latitude. This is depicted in Fig. 2.7 on a Mercator projection. The ogive-shaped curves are the segmented ground tracks for a satellite with a two-hour circular orbit (altitude about 1600 km) and an inclination of 60°. Multiple tracks are shown since the earth rotates counterclockwise as viewed from the north pole and the ground track moves westward. How far west the ground track moves after each orbit depends on the orbital period. For a two-hour orbital period, the ground track moves westward by 30° at the equator (see Fig. 2.7), since the earth's rotation is 15° per hour. There is an additional westward movement due to the fact that the earth is not spherical but oblate (difference in equatorial and polar radii). This produces a precession in the orbit which causes the point where the satellite crosses the equator to occur farther west for an orbit with an inclination of less than 90° (as the one shown in Fig. 2.7). This adds to the equatorial crossing due to the earth's rotation. This phenomenon is referred to as nodal regression, and more will be said in this in a later section.

A single ground track is shown accented with the satellite and footprint at its maximum latitude (60°). It is clear that the footprint extends beyond the satellite ground track.

It may be of further interest to note that a nadir (satellite subpoint)-looking circular beam illuminating the earth will not produce a circular footprint. Further, on a Mercator projection, the footprint will appear

Figure 2.7
Ground tracks of a two-hour circular orbit with 60° inclination, and instantaneous satellite location at 60°N latitude and its sensor footprint. Note the satellite sensor(s) can communicate with a swath of terrain on either side of the ground track.

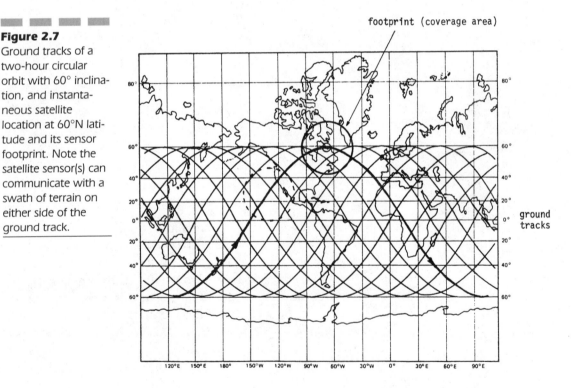

footprint (coverage area)

ground tracks

- Direct circular orbit, i=60°.
- It is noted that in two hours the Earth will have rotated 30°, and therefore the ground track crossing at the equator moves 30° to the west since the earth rotates counter-clockwise as viewed from the north pole.
- Does not reflect nodal regression due to earth's oblateness. This would add to the westward movement due to the earth's rotation.

distorted (at the higher latitudes) because the areas at the higher latitudes will appear stretched out.

In an operational system in which continuous global coverage is required, there normally is a constellation of satellites in several inclined orbits separated in longitude, in addition to several satellites in each plane. An example of this constellation is shown in Fig. 2.8a. The orbits are at the same altitude and inclination and, where possible, equally separated in longitude at the equator.

Another orbital regime are satellites in polar or near-polar orbits. Here the inclination is at 90° or near 90°. An example of this configuration is shown in Fig. 2.8b. Here, too, attempts are made to have equal plane separation, and all satellites are at the same altitude and inclination. The reason for maintaining uniform density is that any perturbations will be witnessed by all the satellites, thus maintaining their uniformity.

2.1.3 Signal Dopplerization

Signals transmitted by satellites will be subjected to dopplerization. This results from motion of the satellite as well as the earth station movement because of the earth's rotation. Consider, for example, the earth station

Figure 2.8
Typical non-GSO satellite constellations used in telecommunications.

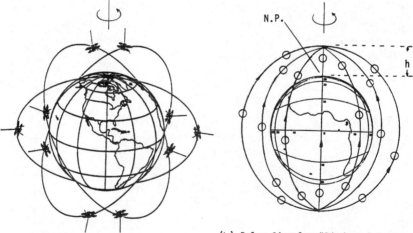

(a) Constellation of satellites in incline orbits.

e.g. • TRW Odyssey system for global communications.

(b) Polar Circular "Birdcage" Orbits for Navigation and Telecommunications.

e.g. • U.S. TRANSIT for global navigation.

• Near-polar orbit of the Motorola IRIDIUM system for global communications.

receiving a signal from the satellite. The received dopplerized signal is given by the expression

$$f_{rec} = f_x \pm f_d = f_x \pm (v_r f_x / c) \qquad [2.1]$$

where f_{rec} = apparent received frequency
f_x = transmitted signal
v_r = relative velocity of the receiver motion with respect to the satellite motion

The ambiguous sign $\pm$ associated with velocity indicates an approaching or ⌃ receding transmitter, respectively. This term is the change in frequency due to the satellite and earth station relative motion. If we assume for the moment that the earth station is fixed and lies on the ground track of a satellite in circular orbit, we have the scenario depicted in Fig. 2.9. For the geometry shown, the earth station is assumed at the horizon and witnessing the maximum Doppler shifted signal from the satellite. When the earth station is at the satellite subpoint or nadir point, $\cos \psi = 0°$ and the Doppler frequency is equal to zero. Clearly this is also true

Figure 2.9
Signal dopplerization
due to spacecraft
orbital motion.

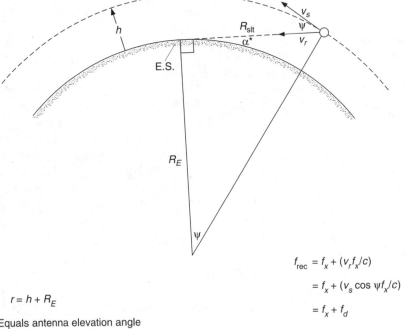

$$f_{rec} = f_x + (v_r f_x / c)$$
$$= f_x + (v_s \cos \psi f_x / c)$$
$$= f_x + f_d$$

$r = h + R_E$

*Equals antenna elevation angle
if earth station is on the ground track.

when the slant range to the ground receiver is minimum in a nonoverflight, or the slant range is normal to the satellite flight path.

For zero elevation angle of the earth station receiver antenna, the slant range and earth radius forms a right angle (see Fig. 2.9). Similarly, the satellite velocity vector forms a right angle with the satellite radius $h + R_E$. This makes the one-half central angle of the earth equal to the depression angle at the satellite. The relative velocity is therefore $v_r = v_s \cos \psi = v_s(R_E/R_E + h)$.

A qualitative plot of the Doppler effect as a function of time is shown in Fig. 2.10. As the satellite approaches the earth station, the Doppler contribution is positive and adds to the transmitter frequency f_x. As the satellite recedes from the earth station, the Doppler frequency subtracts from the transmitted frequency and appears at the earth station as a lower fre-

Figure 2.10
Ogive curve of signal dopplerization.

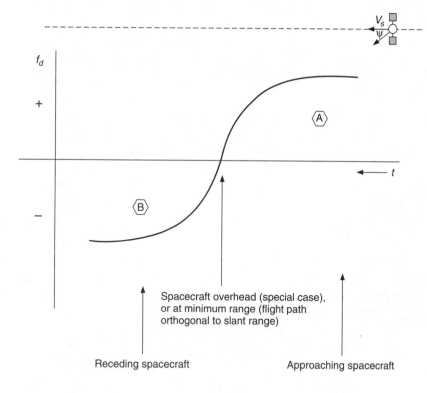

$$f_d = v_r f_x/c = v_{sat} \cos \psi f_x/c$$

- In region Ⓐ received signal is $f_x + f_d$.
- In region Ⓑ received signal is $f_x - f_d$.

quency. Clearly, in the region where the curve crosses the axis, the satellite is overhead (the special case we considered). This is also where the slant range from the satellite to the earth station is shortest.

Now, consider a more realistic case. Since the earth station is in motion due to the earth's rotation, the relative velocity must now include the satellite and earth station velocity. To make this scenario a bit more tractable, assume that the satellite is in a polar orbit and that the earth station is at a latitudinal position west of the satellite orbit. Further assume that the earth station is at a latitude of 30°N. There is no particular significance to this latitude. A sketch representing this scenario is shown in Fig. 2.11.

The angular velocity of the earth station due to the earth's rotation is $\omega = 2\pi/T$, where T is the period for one revolution. If the earth station were on the equator, its velocity would be $v = \omega R_E$. However, being at a higher latitude, its velocity is reduced; we have

$$v_{es} = (\omega R_E) \cos \text{LAT}$$

$$= (2\pi/24)R_E \cos \text{LAT}$$

$$= (\pi/12)R_E \cos \text{LAT} \qquad \text{km/hour} \qquad \text{[2.2]}$$

where LAT = latitude of earth station, 30°N
R_E = earth radius (6378 km at equator)
$2\pi/24$ = earth's angular velocity

Figure 2.11
Scenario of a satellite in polar orbit and an earth station, (for that particular orbit) west of the satellite and in motion due to the earth's rotation.

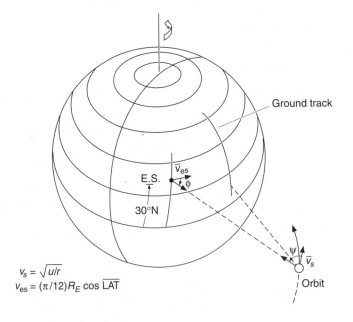

Ground track

$\overline{V}_{es}$

E.S.

ϕ

30°N

ψ

V_s

Orbit

$V_s = \sqrt{u/r}$
$V_{es} = (\pi/12)R_E \cos \overline{\text{LAT}}$

For an earth station at a latitude of 30°N, its tangential velocity due to earth's rotation is therefore

$$v_{es} = 1446 \text{ km/hour} = 401 \text{ m/second} \qquad [2.2']$$

Since the satellite can be seen to the east of the earth station, there will be a closing relative velocity between the earth station and the satellite. This will then combine with the velocity of the satellite with respect to the earth station. For the scenario shown, the composite Doppler frequency will therefore be (these are instantaneous frequencies):

$$f_d = (v_s \cos \psi / c) f_x + (v_{es} \cos \phi / c) f_x \qquad [2.3]$$

where ψ = the angle between the satellite velocity vector and its component in the direction of the earth station

ϕ = the angle between the earth station's tangential velocity vector and its component in the direction of the satellite. Generally not the antenna elevation angle (only if the component is in the vertical coplane with the velocity vector would this be true).

The expression for finding the velocity of the earth station was given previously. The velocity of the satellite in circular orbit is given by

$$v_s = \sqrt{u/r} \qquad [2.4]$$

where $u = GM$ = gravitational parameter for the earth
$= 3.9866 \times 10^5 \text{ km}^3/\text{s}^2$

For example, for an altitude of 1000 km, we have

$$r = h + R_E = 1000 + 6378 = 7378 \text{ km}$$

Therefore, plugging in Eq. [2.4], previously stated, we obtain

$$v_s = 7.3 \text{ km/s} \qquad [2.4']$$

If the earth station were located at the satellite subpoint, the relative velocity of the earth station with respect to the satellite would be zero, and the relative velocity of the satellite velocity with respect to the earth station would also be zero. Therefore, there would be no Doppler. A further example would be if the satellite is at the same latitude as the earth station but east of the earth station, the satellite would contribute zero Doppler, but the earth motion rotating the earth station eastward would make a Doppler contribution. Only in the special case just alluded to, where the earth station is in the satellite nadir beam, will there be zero Doppler contributions from both locations.

Note that as the satellite orbits the earth, it will move closer to the earth station (and of course, will also move west of the earth station). The relative velocity components for both the satellite and the earth station will reduce because of increasing ϕ and ψ, producing a reduced Doppler contribution.

2.1.4 Mechanics of Orbits Used in Mobile Satellite Service

The non-GSO satellites used in mobile satellite systems assume two basic orbital configurations: either circular with varying inclinations or elliptical. *Inclination* is defined as the dihedral angle between the equatorial plane and the plane of the orbit. The earth's center always lies in the plane of the orbit. The orbit inclination may vary from 0° (in the equatorial plane) to 90° (in the polar plane) or even beyond 90° for special applications. Orbits with inclinations between 0° and 90° are referred to as direct or prograde orbits. Orbits at 90° or close to 90° inclinations are dubbed polar orbits. Orbits with inclinations greater than 90° are referred to as retrograde orbits. Note that in this orbit the spacecraft orbits in the opposite direction to the rotation of the earth.

A satellite in orbit can be defined by six Keplerian elements. These are illustrated in Figs. 2.12 and 2.13. Elliptical orbits have one of their foci at the center of the earth, and the other focus is referred to as the vacant focus. Orbits providing service to the northern hemisphere are oriented to have their farthest distance, or apogee, in the northern hemisphere and the perigee (closest point to the earth) in the southern hemisphere. By Kepler's second planetary laws of motion, the satellite velocity is slower in the apogee region than in the perigee region. To an observer or an antenna in the northern hemisphere, the satellite appears to dwell in proximity to the apogee region. Since this region is the farthest from the earth, antennas viewing the satellite will have high elevation angles for long periods of time because the satellite spends a large part of its time in this region. This is represented in Fig. 2.14. Typically, the high elevation viewing times are in the order of 6 to 8 h, for an orbit period of 12 h.

It is clear that the distance to satellites in elliptical orbits will vary and the satellite may even go out of sight as it goes into the perigee region of the orbit. Therefore, a satellite beam or beams illuminating the earth will have varying footprint sizes for a fixed antenna beam. This can be com-

Figure 2.12

Six keplerian orbital elements are required to specify a satellite orbit completely.

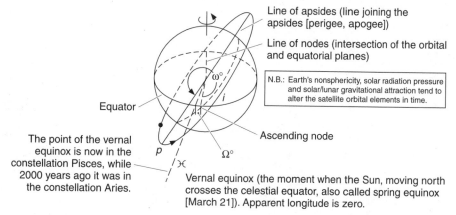

Line of apsides (line joining the apsides [perigee, apogee])

Line of nodes (intersection of the orbital and equatorial planes)

N.B.: Earth's nonsphericity, solar radiation pressure and solar/lunar gravitational attraction tend to alter the satellite orbital elements in time.

Equator

Ascending node

The point of the vernal equinox is now in the constellation Pisces, while 2000 years ago it was in the constellation Aries.

Vernal equinox (the moment when the Sun, moving north crosses the celestial equator, also called spring equinox [March 21]). Apparent longitude is zero.

Three define the orbit plane orientation

$\Omega°$: Right ascension of ascending node, measured in plane of equator from the direction of the vernal equinox* to the direction of the ascending node, or the intersection of the orbit plane with the equatorial plane ($0° \leq \Omega \leq 360°$).

$i°$: Inclination of orbit plane, dihedral angle between plane of equator and plane of orbit ($0° \leq i \leq 180°$)

$\omega°$: Argument of perigee, angle between the direction of ascending node and direction of perigee. ω is in the plane of the orbit ($0° \leq \omega < 360°$).

* Vernal equinox is the ascending node of the earth's orbit around the sun.

Three define parameters of orbit

e: Eccentricity ($0 \leq e < 1$) [closed orbits].

a: Semimajor axis of orbit.

τ: Time of perigee passage, or position of body for anytime τ.

a, e: Determine geometry or shape of orbit.

i, ω, Ω: Gives orientation of orbit.

pensated for by reducing the satellite antenna beam width as the vehicle moves farther from the earth. In addition, it must be continuously steered in order to provide service to a particular area when that area is in view. Since the satellite goes out of view for part of its orbit, at least three must be provided in the plane to obtain continuous coverage, with provisions for handover from one satellite to the next. For elliptical orbits, the inclination is an important parameter. Because of the oblateness of the earth, the inclination angle must be maintained at particular values to prevent anomalies in the orbital plane. This subject will be discussed further in a later section.

The Russians have extensively used elliptical orbits to provide service to their territory (since a large part of Russia lies in the northern latitudes).

Figure 2.13
Geometry of elliptical
orbital parameters.

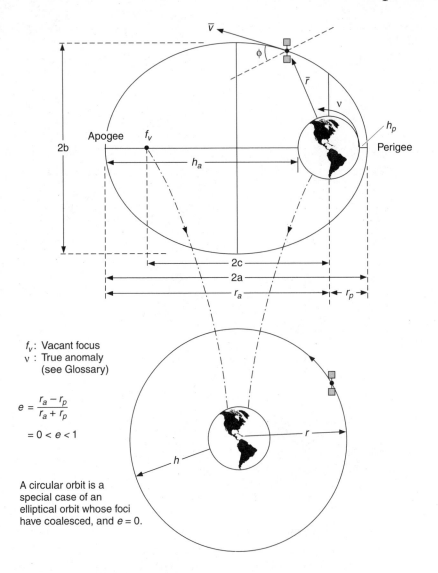

Figure 2.13
Geometry of elliptical
orbital parameters.

f_v: Vacant focus
v : True anomaly
(see Glossary)

$$e = \frac{r_a - r_p}{r_a + r_p}$$

$$= 0 < e < 1$$

A circular orbit is a
special case of an
elliptical orbit whose foci
have coalesced, and $e = 0$.

The apogee in the northern hemisphere provides long periods of time
when the satellite(s) appear to dwell over their territory and afford high
elevation angles for their earth stations.

The apogee altitude for their satellites is about 40,000 km and perigee
(in the southern hemisphere) is about 1000 km. The eccentricity (see Fig.
2.13) is about 0.7. The period is 12 h, therefore producing two orbits per
day. The inclination is critical and is 63.4°. More will be said about this
angle in a later section.

Figure 2.14

Highly elliptical orbit:

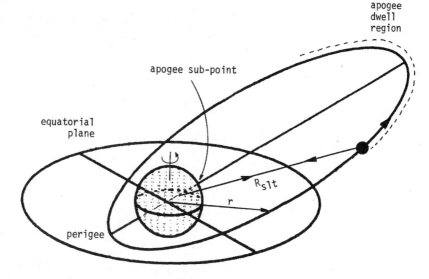

- Orbital period: $2\pi a^{3/2}/\sqrt{u}$
- $r_p = a(1-e)$
- $r_a = a(1+e)$
- $v = \sqrt{GM[(2/r)-(1/a)]}$
- $v_p^2 = ur_a/ar_p$, $u = GM$
- $v_a^2 = ur_p/ar_a$
- mean velocity = $\sqrt{u/a}$

Not to scale

The velocity of a satellite in elliptical orbit varies depending on its location and on its distance from the earth. At perigee it is traveling at its greatest velocity (Kepler's second law) (see Fig. 2.14). At apogee it is traveling at its slowest velocity. The velocity at any range can be given by the *vis viva** energy equation:

$$v^2 = G(M + m)[(2/r) - (1/a)] \qquad [2.5]$$

where $GM = u$ = gravitational parameter, 3.986×10^5 km³/s²
G = Newton's gravitational constant

---

*An equation that crops up in the solution of the two-body problem in celestial mechanics.

M = mass of the earth
m = mass of space vehicle ($m \ll M$), may be neglected
r = satellite distance from center of earth
a = ellipse's semimajor axis = $(\frac{1}{2})(r_p + r_a)$
r_a = apogee radius = $a(1 + e)$
r_p = perigee radius = $a(1 - e)$
e = eccentricity ($0 < e < 1$)

The velocity at perigee is given by

$$v_p = \sqrt{u/p}(1 + e) = \sqrt{u/a}\sqrt{(1 + e)/(1 - e)} \qquad [2.6]$$

where p = latus rectum of the ellipse and is equal to $a(1 - e^2)$
The velocity at apogee is given by

$$v_a = \sqrt{u/p}(1 - e) = \sqrt{u/a}\sqrt{(1 - e)(1 + e)} \qquad [2.7]$$

Since circular orbits are a special case of an elliptical orbit in which the foci coalesce, the eccentricity is zero and $r = a$. The velocity expression thus simplifies to

$$v_c = \sqrt{u/r} \qquad [2.8]$$

where r is now the constant orbital radius, and the velocity is constant for a circular orbit but changes with altitude.

For the circular orbit, the velocity may also be found by dividing the distance around the orbit by the time to negotiate the orbit. That is,

$$v_c = 2\pi r/T = 2\pi(R_E + h)/T \qquad [2.9]$$

where h = altitude
T = period of orbit
r = orbital radius
R_E = earth's radius

The orbital period of a circular orbit is given as

$$T = 2\pi\sqrt{r^3/u} \qquad [2.10]$$

The period of a satellite in elliptical orbit is given by the same formula as for a circular orbit, provided the distance r from the earth's center, which varies for the elliptical orbit, is replaced by the average distance a from the earth's center. That is, halfway between the maximum distance from the earth's center $r_{max} = r_a$ and minimum distance $r_{min} = r_p$. Therefore, a is actually the semimajor axis or

$$a = (\tfrac{1}{2})(r_a + r_p) \qquad [2.11]$$

The period is therefore equal to

$$T = 2\pi\sqrt{a^3/u} = 2\pi a\sqrt{a/u} \qquad [2.12]$$

Curves for period and velocity for circular orbits are given in Figs. 2.15 and 2.16. Even though Fig. 2.14 applies to circular orbits, it may also be used for elliptical orbits if r is now replaced by a, the average of the minimum and maximum orbital radius as alluded to previously.

Typical velocities for low earth circular orbits ($h = 1000$ km) are about 7.3 km/s (16,300 mi/h) and periods in the order of 105 minutes. For satellites in medium earth orbit ($h = 10,000$ km), the velocity is about 5 km/s (11,200 mi/h) and a period of about 6 h. By comparison, for a satellite in GSO ($h = 35,800$ km), the velocity is about 3 km/s (6700 mi/h), and the period is about 24 h.

2.1.5 Multiple Satellites

As indicated previously, low earth orbit satellites operating singularly are not able to provide continuous and global service. Multiple satellites must be launched into several orbits. The total number depends on the altitude

Figure 2.15
Orbital period as a function of satellite altitude.

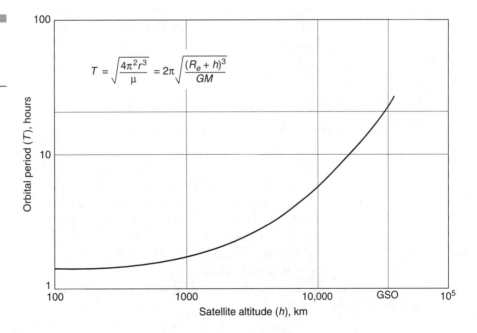

$$T = \sqrt{\frac{4\pi^2 r^3}{\mu}} = 2\pi\sqrt{\frac{(R_e + h)^3}{GM}}$$

Orbital period (T), hours

Satellite altitude (h), km

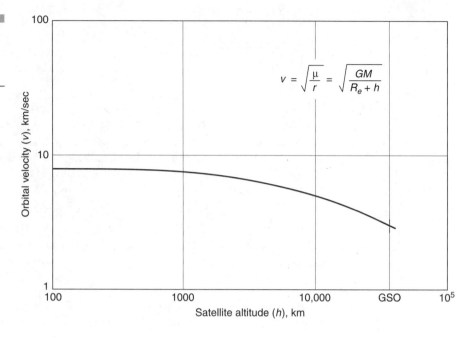

and the required antenna elevation angle of the terrestrial transceivers. The latter requirement is to provide reliable service in the presence of multipath signals. They may assume any direct orbit inclinations between 0° and 90°, but all must have the same inclination and altitude. This is to ensure that they remain fixed in the same orientation with respect to each other in the presence of the earth perturbations. The subject of perturbations will be discussed in the next section.

2.2 Orbit Departure from Ideal Keplerian Laws

A satellite orbiting the earth would ideally adhere to the keplerian laws of planetary motion if there were no perturbing forces influencing its orbit. Some of these forces are external to the earth, but others are indigenous to the earth. In particular, the earth is not a perfect sphere but is shaped more like an ellipsoid. The difference in length of the polar radius and the equatorial radius makes the earth oblate,* and this radius difference produces several deleterious effects in the satellite orbit. Two salient anomalies are the regression or advancement of the line of nodes or right

---

* The equatorial diameter is greater than the axial diameter by about 21 km.

ascension of the ascending node (RAAN, see Fig. 2.12) and the rotation of the lines of apsides or the line joining the apogee and perigee (major axis of ellipse) of the elliptical orbit (see Fig. 2.12). For circular orbits, the line of apsides has no significance since a major axis is undefined.

2.2.1 Right Ascension of Ascending Node

The nodal regression (or advancement) due to the earth's oblateness parameter J_2^* is given by the following relationship. J_2 is a dimensionless number that characterizes the departure of the earth from a true sphere. The migration of the right ascension of the ascending node (RAAN) at any time t is given by

$$\Omega = \Omega_0 - \frac{3nJ_2 R_E^2}{p^2} \cos i(t - t_0) \qquad \text{degrees per day}$$

$$= \Omega_0 - (d\Omega/dt)(t - t_0) \qquad\qquad\qquad \text{[2.13]}$$

where n = the mean orbital motion or average angular rate of the orbit and is equal to $2\pi/T$, where T is the period of orbit
$\quad J_2$ = second harmonic oblateness coefficient; for earth has a value 0.001826
$\quad R_E$ = earth equatorial radius (6378 km)
$\quad i$ = orbit inclination
$\quad p = a(1 - e^2)$
$\quad a$ = semimajor axis
$\quad e$ = eccentricity

At the epoch point when $t = t_0$, the orbital precession is $\Omega = \Omega_0$, the value of Ω at epoch. The minus sign prefix indicates that the nodal crossing regresses for each orbital revolution and subtracts from the epoch value Ω_0 (see Fig. 2.17a). When the inclination is between 90° and 180°, the precession is positive (advancement) in the direction of earth's rotation. Each rotation adds to the epoch Ω_0 (see Fig. 2.17b).

In Eq. [2.13], if mean motion n is replaced by 2π, we have the variations in radians per revolution since $nT = 2\pi$. In addition, if we replace p with $a(1 + e^2)$, we obtain the rate of change of precession.

---

* J_2: One of the zonal harmonic coefficients of the earth's gravitational potential and a major cause of perturbation influencing the argument of perigee (rotation of line of apsides) and migration of the ascending node. Gravitational potential is interesting per se, but not within the scope of this book.

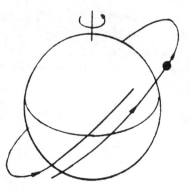

Line of nodes regresses
westward if orbit is
"direct" (i.e., 0<i<90°)

• for i=90°, no regression
• amount of regression is a
function of inclination &
altitude

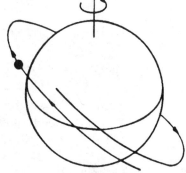

Line of nodes advances *
eastward if orbit is
retrograde (90°<i<180°)

* In direction of earth's rotation.

$$\dot{\Omega} = d\Omega/dt = -3\pi(R_E/a)^2 J_2(1 - e^2)^{-2} \cos i \qquad \text{radians/orbit} \qquad [2.14]$$

or

$$= -9.964(1 - e^2)^{-2}(R_E/a)^{3.5} \cos i \qquad \text{degrees/day} \qquad [2.14']$$

Equation [2.14] also applies to circular orbits where a is replaced by r and $e = 0$.

We may also find the precession regression (or advancement) per orbit from Eq. [2.14']. There are 1440 minutes in a day. If the period is T, the number of revolutions per day is $1440/T = N$. Since we know the precession rate per day from Eq. [2.14'], we have $d\Omega/dt/\text{day}/N$ degrees per orbit.

The minus sign in the equation above, Eq. 2.14', indicates that the node drifts westward $(0 < i < 90°)$. There is no nodal migration for $i = 90°$.

A manifestation of the nodal regression/advance phenomenon is depicted in Fig. 2.17. In Fig. 2.17a, the inclination angle is less than 90°, and the node (equatorial crossing) moves westward. In Fig. 2.17b, for inclinations greater than 90°, there is an advance in the nodal crossing in the direction of the earth's rotation. In Eq. [2.14], since cosine i is $90° < i < 180°$, the sign of $\dot{\Omega}$ changes from minus to plus.

The nodal regression/advancement is also a function of orbital altitude. This is shown in Fig. 2.18. As an example of this nodal regression for some practical systems, the U.S. shuttle with a parking orbit altitude of 300 km (period = 90 min) and inclination angle of 28.5° has a nodal regression of −7.5°/day (a rather significant perturbation). For the GPS navigational system, operating at an altitude of 20,000 km and inclination of 55°, the

Figure 2.18
Nodal regression rate for a circular orbit as a function of inclination.

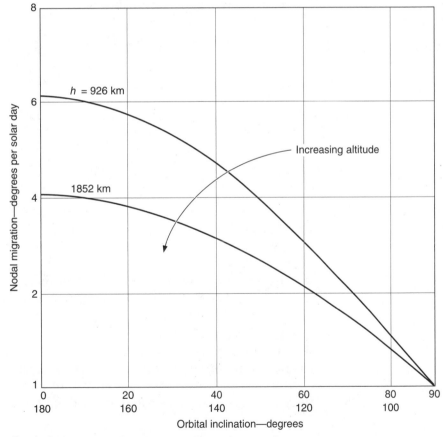

Epoch: Ω_0

• For direct orbit ($0 < i < 90°$): westward nodal migration.
• For retrograde orbit ($90° < i < 180°$): eastward nodal migration.
• For strictly polar orbit ($i = 90°$): there is no migration.

Figure 2.19
Nodal migration as a
function of inclina-
tion, with parametric
orbital period in min-
utes.

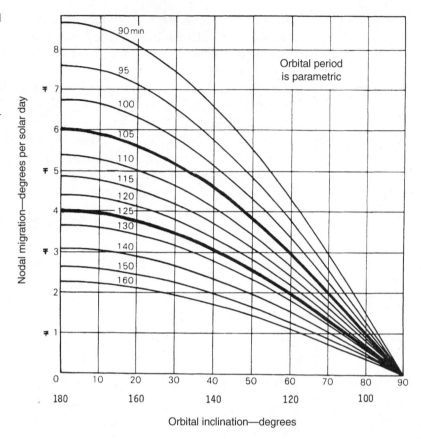

nodal rate change is −0.033°/day. For a satellite at GSO, the rate of change
is 0.013°/day.

Figure 2.19 shows the nodal migration in a slightly different form. Here
the migration is given in terms of orbital period. This tracks the data
shown in Fig. 2.18, and clearly the nodal precession decreases with increasing altitude or orbital period.

The two accented curves ($T = 105$, $T = 125$ min) correspond roughly to
the two curves given in Fig. 2.18. The uppermost curve in Fig. 2.19 would
correspond to the U.S. shuttle ($h = 300$ km) with an inclination of 28.5°.

In summary, when the rate of change of nodal precession is negative,
there is a regression westward.[*] This corresponds to an orbital inclination

---

[*] Referring to Eq. (13), some authors [Escobal, Roy] prefix the second term by a plus sign.
Therefore, when $i < 90°$, the plus refers to westward migration. For $i > 90°$, the sign is negative, therefore indicating an eastward movement.

less than 90°. When the rate of change is positive, the precession is eastward and corresponds to an inclination greater than 90°.

2.2.2 The Earth Keeps Spinning

In addition to the nodal migration because of the earth's perturbations, there is also the earth's rotation which impacts the nodal crossings. A satellite crossing the equator in a northerly or ascending direction will see a change in longitude for each successive crossing. The earth rotates through one revolution in a period T_E which is about 24 h. If the satellite period is T_s, the earth's rotation in one orbital rotation of the satellite is

$$\Delta\theta = -2\pi(T_s/T_E) = -2\pi(T_s/24) \qquad [2.15]$$

For example, if $T_s = 1$ hour, therefore

$$\Delta\theta = -0.262 \text{ radians per hour}$$

$$= -15° \text{ per hour}$$

Therefore, the total longitudinal migration of the satellite node for a direct orbit $(0 < i < 90°)$ is

$$\Delta\theta = \Delta\theta + \Delta\Omega \qquad \text{radians per satellite orbit} \qquad [2.16]$$

where $\Delta\Omega$ = the nodal regression

If the node regresses, $\Delta\theta$ is greater than 15° (for a 1-hour orbit). For a retrograde orbit, the $\Delta\Omega$ is subtracted from $\Delta\theta$. This can be observed by referring to Fig. 2.17.

2.2.3 Sun-Synchronous Orbit

Generally, orbits for telecom satellite-based wireless communications use inclinations less than 90°. However, circular orbits with inclinations greater than 90° are also used for special applications, and one is referred to as sunsynchronous since the orbit follows the sun.* Recalling from previous

---

* An application has been submitted to the FCC for a non-GSO (LEO) satellite using the sun-synchronous mode with the orbit in the twilight zone. That is, the orbit is oriented perpendicular to the earth-sun line. The satellite will therefore always be in sunlight. This is a data-only system.

information, the orbit plane precesses eastward and at the same rate as the earth revolves around the sun, which is also eastward. The satellite will be at the same latitude at the same time each day but not at the same longitude because the earth is rotating under the satellite. However, the satellite will return to the same longitude after roughly 17 to 18 days.

Reference to Fig. 2.20 illustrates the orbital movement with respect to the sun. It is noted that as the earth rotates counterclockwise, viewed from the north pole, the orbit also precesses in the same direction. By the

Figure 2.20
Graphics of sun-synchronous operation.

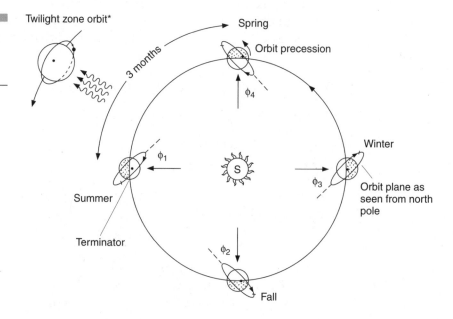

- The orbit plane that precesses 360° per annum is called sun-synchronous. For precession of 360° per annum, the precession rate is:
      (360°/365.25 days) = 0.986°/day east
- Amount of orbital plane precession is a function of satellite altitude (or radius), inclination and eccentricity:
      $(d\Omega/dt) = -9.98(R_E/a)^{3.5}(1 - e^2)^{-2} \cos i$, deg/day
  For circular orbit $a = r$, $e = 0$. Minus sign signifies counterclockwise rotation of orbital plane as seen from north pole. Note: $\cos i$ is negative.
- Passes over the same part of earth the same time each day (crosses a given latitude at the same local time).
- The sun-synchronous orbit is retrograde.

      9.98 incorporates several fixed quantities.

* It is noted for the satellite orbit operating in the twilight zone, it will be above the terminator line, or the demarcation between day and night. The plane of the orbit maintains a fixed orientation with respect to the sun.

proper choice of orbit altitude, eccentricity, and inclination, the orbit plane can be made to change at the same rate as the earth circles the sun. That is, at the rate of 0.986°/day eastward or on an annual basis yields 0.986 × 360 days = 365.25 days.

In Fig. 2.20, the angle ϕ_n is maintained constant throughout the year. The reader will recognize that this movement is the nodal advancement and is given by the equation in the second bullet on Fig. 2.20. Note that the minus sign prefixed before the equation coupled with the cosine factor of the inclination angle produces a plus precession indicating that the orbit nodal crossover is in the *same* direction as the earth's rotation (i.e., an advancement of the nodal crossing).

Solving Eq. [2.14'] for i and plugging in known values ($e = 0$, $a = h + R_E$) we obtain

$$i = \arccos -0.99[(R_E + h)/R_E^{3.5}] \qquad [2.17]$$

The minus sign prefix implies that the orbit is retrograde ($90° < i < 180°$). This relationship is shown plotted in Fig. 2.21.

2.2.4 Argument of Perigee

In addition to the nodal precession anomaly caused by the earth's oblateness, there is the rotation of the line of apsides or rotation of the argument (angle) of perigee. This perturbation is more directly relatable to elliptical orbits, since for circular orbits apsides have little significance.

The argument of perigee at any time t is given by the expression

$$\omega = \omega_0 - \frac{3nJ_2R_E^2}{2a^2(1-e^2)^2} [(5/2) \sin^2 i - 2](t - t_0)$$

$$= \omega_0 - (d\omega/dt)(t - t_0) \qquad [2.18]$$

where, at $t = t_0$, the epoch time $\omega = \omega_0$, the initial value (at epoch) of ω.

The second term

$$-\frac{3nJ_2R_E^2}{2a^2(1-e^2)^2} [(5/2) \sin^2 i - 2](t - t_0) \qquad [2.19]$$

is the rate of change of the argument of perigee in radians per revolution ($n = 2\pi/T$ radians/period). All other terms have been defined previously.

The rate of change of the argument of perigee may also be given in degrees per day

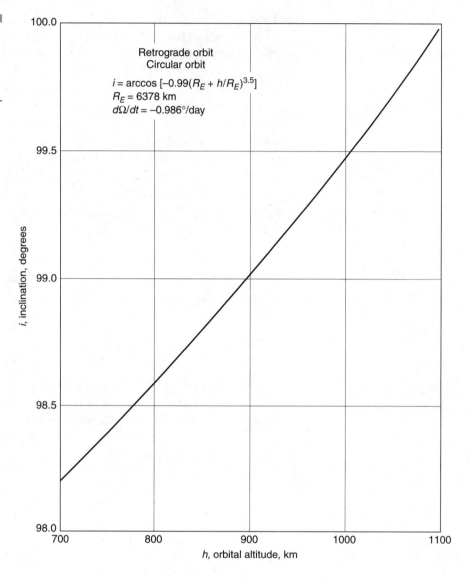

$$\frac{d\omega}{dt} = 5(R_E/a)^{3.5}(5 \cos^2 i - 1)(1 - e^2)^{-2} \qquad \text{degree/day} \qquad [2.20]$$

There are two orbit inclination values which yield no rotation of the line of apsides. If we let $d\omega/dt = 0$ in Eq. [2.20], and we assume a, R_E, and e are fixed (logical), we have

$$5 \cos^2 i - 1 = 0 \qquad [2.21]$$

Or if one prefers, one may use Eq. [2.5] and we obtain

$$(5/2) \sin^2 i - 2 = 0 \cdots \qquad\qquad [2.22]$$

since Eqs. [2.21] and [2.22] are identical. Both equations give values of i which are $i = 63.4°$ and $116.6°$. These are referred to as critical angles. These angles are independent of a or e (thus allowing flexibility for the system designer).

A plot of the apsidal rotation as a function of inclination angle, with parametric apogee altitudes and fixed perigee altitude, is shown in Fig. 2.22 for a moderate elliptical orbit (lowercase e). The rate of apsidal rotation decreases with an increasing semimajor axis which is usually accompanied by an increase in eccentricity. The apsidal rotation is zero at the two critical angles shown on the curve.

Figure 2.22
Rotation of the line of apsides for an elliptical orbit.

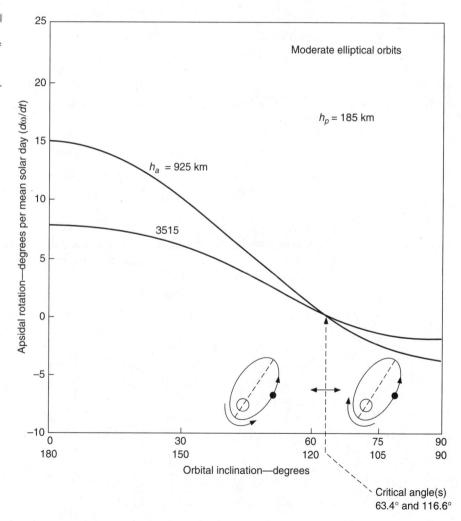

If i is not one of the critical angles, the sign of $d\omega/dt$ may be either plus or minus depending on whether the inclination angle is above or below the critical angles. It should be made clear that the orbit inclination remains *constant,* but the orbit rotates in its plane at the fixed inclination. For example, for an orbit with i *less* than 63.4° (or >116°), the line of apsides will rotate in the same direction as the satellite motion. For an orbital inclination greater than 63.4° (or <116°) the line of apsides will rotate in the opposite direction (reduces ω) to the satellite motion. Both of these situations are illustrated in Figs. 2.22 and 2.23. In both displays, the apogee is assumed in the northern hemisphere. In addition, there is migration of the nodal crossing (not shown in Fig. 2.22). In both cases in Fig. 2.23, the ascending nodal crossing will manifest a westward regression.[*] As indicated in previous sections, the westerly regression will add to the satellite's longitudinal movement west.

The orbital apsidal rotation and precession are further amplified in Fig. 2.24. Here movements for the 116° critical angle are also displayed. In this representation the orbits are shown normal to the plane of the page as opposed to the isometric in Fig. 2.23. Apogee is at the upper end of the orbit.

A summary can be gleaned from Eq. [2.18][†] as follows: when $d\omega/dt$ is positive, $i < 63.4°$, the argument of the perigee ω rotates in the same direction as the satellite. The argument decreases when $d\omega/dt$ is negative, $i > 63.4°$; the argument of the perigee rotates in the opposite direction from the satellite motion. These are clearly demonstrated in Fig. 2.23. It is necessary to control the inclination angle to prevent apsidal rotation, which can move the apogee out of the northern hemisphere. As indicated previously, the apsidal rotation rate will be small per day for large values of the semimajor axis a.

An application of this orbit property in which the apsidal rotation is of concern is central in the Russian Molniya-type orbit. This is an elliptical orbit with $e = 0.7$ and a period of 12 h. The apogee distance is about 40,000 km and resides in the northern hemisphere. Actually because of the 12-hour orbits, two apogees will exist in a 24-hour period. The other apogee is over North America, or at a longitudinal distance of 180°. The perigees are in the southern hemisphere at an altitude of about 1000 km, and the argument of the perigee is about 270°. The high apogee distances in the northern latitudes provide high antenna elevation angles for users in these regions.

---

[*] The nodes move in a direction opposite to the direction of the satellite and hence regression of the nodes.

[†] Depends on the sign of $(5 \cos^2 i - 1)$.

Figure 2.23
Direction and change of the argument of the perigee (ω) as a function of the orbital inclination in reference to the critical angle (i').

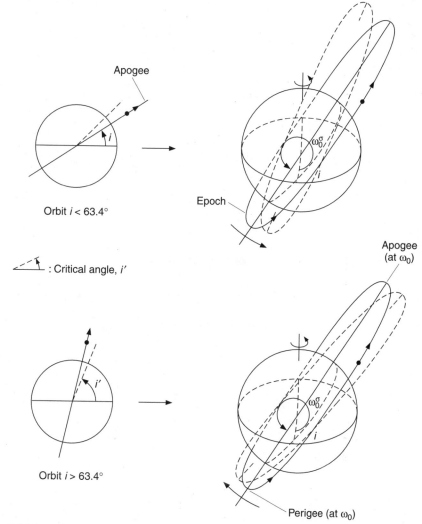

Apogee

Orbit *i* < 63.4°

: Critical angle, *i'*

Epoch

Apogee (at ω₀)

Orbit *i* > 63.4°

Perigee (at ω₀)

- Orbit *i* remains constant.
- In either case, the apogee will drift into the southern hemisphere eventually.

   At the high apogee altitude, and because of Kepler's law of planetary motion, the satellite appears to the ground user to move slowly or dwell around the apogee altitudes. This has been referred to as the apogee dwell (see Fig. 2.14). This apparent stationarity of the satellite is about 6 to 8 h, with high elevation angles of the earth station antenna.

Figure 2.24
Rotation of the line of apsides and the precession of the line of nodes for elliptical orbits.

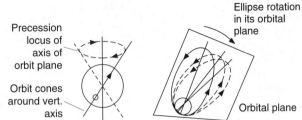

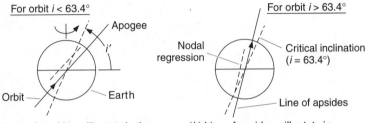

For direct motion orbits

For orbit $i < 63.4°$

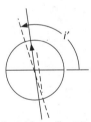

(*a*) Line of apsides will rotate in the same direction of satellite motion, in this case apogee rotates into page. Westward nodal regression.

For orbit $i > 63.4°$

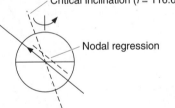

(*b*) Line of apsides will rotate in opposite direction of satellite motion, that is, apogee out of page. Westward nodal regression.

For retrograde motion orbits

For orbit $i < 116.6°$

(*c*) Line of apsides will rotate in opposite direction of satellite motion, that is, rotation of apogee out of page. Eastward nodal advancement.

For orbit $i > 116.6°$

(*d*) Line of apsides will rotate in same direction of satellite motion, that is, rotation of orbit apogee into page. Eastward nodal rotation.

• No precession for $i = 90°$

Returning to the apsidal rotation, and considering the Molniya parameters ($a = 26{,}600$ km, $e = 0.70$, and $i = 63.4°$), Eq. [2.20] reduces to

$$d\omega/dt = 0.1204(5 \cos^2 i - 1)$$
$$= 0.1204[(\tfrac{5}{2})(1 + \cos^2 i) - 1] \qquad [2.20']$$

This is plotted in Fig. 2.25. We observe the small apsidal rotation rates. The rotation rates are substantially smaller than those shown in Fig. 2.22, making it less sensitive to perturbing forces which would rotate the orbit in its plane. As observed in previous sections, apsidal rotation is positive, or in the same direction of satellite rotation, for orbit inclination angles less than 63.4°. It is negative for orbit inclination greater than 63.4°.

A satellite orbit of small eccentricity can, in a matter of days (see Fig. 2.22), reverse its apogee and perigee positions, if disturbances are not corrected by onboard thrusters. This phenomenon actually occurred for the passive communication satellite (balloon) in the early 1960s. The orbit was slightly elliptical with inclinations of 47° (Echo I) and 81° (Echo II). The

Figure 2.25

Apsidal rotation for elliptical Molniya-type orbits.

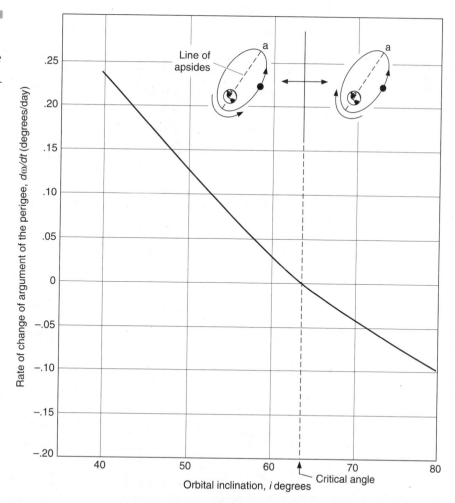

nominal altitudes were in the range 1300 to 1700 km. Both a and e were small (therefore, $d\omega/dt$ high). Both the perigee and apogee altitudes varied, and the two reversed positions every few days. Here the solar radiation pressure was instrumental in causing this reversal.

In summary, the orbital perturbations caused by the oblateness of the earth are indicated in the following list.

- If $i = 63.4°$ or $116.6°$, there is no apsidal rotation, but there is nodal regression.
- When $d\omega/dt$ is positive ($i < 63.4$ or $i > 116.6°$), argument of perigee rotation is in the same direction as the spacecraft motion.
- When $d\omega/dt$ is negative ($i > 63.4°$ or $i < 116.6°$), argument of perigee rotation is in the opposite direction of spacecraft motion.
- There is no nodal regression for $i = 90°$, but there is apsidal rotation.
- Note, if the earth were perfectly spherical, the J_2 factor $= 0$, and no apsidal rotation or nodal regression would occur.
- Both perturbations are a function of semimajor axis, eccentricity, and mean motion.
- Rotation of the line of apsides for a circular orbit has no meaning, but there is nodal regression (for $i \neq 90°$).

2.3 List of Symbols and Variables

a	Semimajor axis.
G	Newton's gravitational constant.
GM	Gravitational parameter.
h_a	Apogee altitude.
h_p	Perigee altitude.
i	Orbital inclination.
J_2	Second zonal harmonic of the earth's gravitational potential function.
M	Mass of the earth.
n	The mean angular rate of a satellite is given by the mean motion, $n = 2\pi/T = \sqrt{u/a^3}$.
r_a	Apogee distance (from center of earth).
r_p	Perigee distance (from center of earth).
R_{slt}	Slant range (to satellite).
ω	Argument (angle) of perigee (measured from ascending node).
Ω	Longitude of ascending node (with respect to the point of Aries).

References

Bate, R. R., et al., "Fundamentals of Astrodynamics," Dover Publ., Inc., New York, 1971.

Battin, R. H., "The Mathematics and Methods of Astrodynamics," AISS Education Series, AIAA, Washington, D.C., 1968.

Blitzer, L., et al., "Perturbations of a Satellite Orbit Due to the Earth's Oblateness," *J. of Applied Physics,* October 1956.

Burington, R. S., *Handbook of Mathematical Tables and Formulas,* Handbook Publishers, Inc., Sandusky, Ohio, 1958.

Chobotov, V. A., (ed.), "Orbital Mechanics," AIAA Education Series, AIAA, Washington, D.C., 1991.

Escobal, P. R., "Methods of Orbit Determination," J. Wiley & Sons, New York, 1976.

Pattan, B., "Satellite Systems—Principles and Technologies," Van Nostrand Reinhold, New York, 1994.

Roy, A. E., "The Foundations of Astrodynamics," Macmillan Co., New York, 1965.

Wiesel, W. E., *Spacecraft Dynamics,* McGraw-Hill Book Company, New York, 1989.

Satellite-Based
Cellular
Communications

3.1 Introduction

3.1.1 The Present

Satellite-based mobile communications have been going through an evolutionary change in the past 10 years, starting with an Inmarsat-type of mobile communications with the satellites in GSO where initially global beams are used to provide service to ships at sea. In 1996, Inmarsat launched two (of five) Inmarsat 3 satellites which produced global spot beams where the earth's disk is divided into large coverage areas serviced by individual spot beams. For the same satellite-transmitted power, the spot beams provide considerably greater effective isotropic radiated power (EIRP) than global beams. This clearly relaxes the design burden on the ground terminal since it sees a larger satellite antenna gain divided by system noise temperature (G/T_s) (larger satellite receive antenna) and a greater downlink EIRP. It is claimed that the higher power-flux density incident on the surface of the earth will allow notebook-size earth terminals.

This era was followed by satellites in GSO providing several spot beam—type services to terrestrial mobile units, either in vehicles or suitcase-size earth terminals. With the reasonably high EIRP laid down by the satellite, the mobiles can use medium-gain antennas (directional) for both data reception and voice service. However, it is not able to supply service to handheld transceivers.

Providing service to handheld transceivers from satellites in GSO will require a very large unfurlable antenna (high-gain) to fit into the shroud of a launch vehicle for operation in the lower microwave region and complementary transmitter power. For example, at L-band (1 to 2 GHz), the antenna size may be 10 to 15 m in diameter. The reason for this requirement is clear. The handheld transceiver is a very poor performance structure with power output in the order of tens of milliwatts and antenna gains in the order of 0 to 3 dB. The amount of power radiated by the handset is limited by the battery power (and its weight), but more importantly by the human safety concern. This problem is still being addressed. The ground segment therefore requires a large power-flux density incident on the antenna (resulting from a high satellite EIRP) and a large G/T at the satellite (high-gain satellite receive antenna) to intercept the small signal radiated from the transceiver (handheld). These comments can be demonstrated by the link equations depicted in Fig. 3.1. Note the dotted-boxed terms. On the uplink, the earth station (transceiver) EIRP is small, and the satellite figure of merit (G/T) needs to be large. On the downlink, the earth station figure of merit is small, and the satellite EIRP needs to be large. Most of the EIRP ($P_t \times G(\theta, \phi)$, where P_t is the satellite transmit power and $G(\theta, \phi)$ is the gain of its antenna in the direction of the earth station) from the satellite comes from a high-gain antenna.

The technology for this kind of service from GSO is not here yet (as of 1996), even though it is on the drawing boards.*

At the present time, an organization which can supply mobile service to suitcase-size transceivers (from GSO) is American Mobile Satellite Corporation (AMSC) using a GSO satellite located at 101°W. The satellite provides service at L-band for user communications and uses the Ku-band (11 to 18 GHz) for interfacing with gateways which connect to the public switched telephone network (PSTN).

All of the GSO mobile satellites providing voice service depend on a directional beam ($G > 10$ dB) earth station antenna. Lower gain antennas may be used, but provide only low data rates and paging service (no voice).

---

* Both Hughes and Lockheed-Martin are developing this technology.

Figure 3.1

Link equations showing the requirements placed on an MSS satellite in GSO to provide service to handheld transceivers.

$P_r = P_t \, G_t \, G_{rec} \lambda^2/(4\pi R_{slt})^2 M$ Friis equation

$N = kT_0 B \cdot \overline{NF}$

$N_0 = kT_s$ (includes noise figure & antenna noise temp.)

UP:

$(C/N_0)_{up} = P_t G_t G_r \lambda^2/(4\pi R)^2 kT_s L$

$(C/N_0)_{dB} = \underbrace{10 \log P_t G_t}_{\text{Earth station EIRP}} - 20 \log (4\pi R/\lambda) + \underbrace{10 \log (G_{rs}/T_s)}_{\text{Satellite figure of merit}} - 10 \log k - 10 \log L - M$

DWN:

$(C/N_0)_{dwn} = \underbrace{10 \log P_s G_s}_{\text{Satellite EIRP}} - 20 \log (4\pi R/\lambda) + \underbrace{10 \log (G_r/T_r)}_{\text{Earth station figure merit}} - 10 \log L - 10 \log k - M$

Power flux density (PFD) at satellite or earth station: $\phi = P_t G_t (\theta, \phi)/4\pi R^2$ watts/m^2

Antenna gain: $4\pi A_p \eta/\lambda^2$, $A_{physical}\eta = A_{eff}$, effective area

$P_{received} = \phi A_{eff}(\theta, \phi) = [P_t G_t (\theta, \phi)/4\pi R^2]A_e(\theta, \phi)$

A_p: Physical aperture area (aperture antennas, not endfire or wire antennas)

N: Receiver noise power

N_0: kT_s noise power density

M: Margin power above the minimum requirements to ensure performance against any propagation anomalies

R_{slt}: Slant range from es/sat or sat/es

L: Additional losses (e.g., polarization, pointing, …)

η: Antenna efficiency

The next phase in mobile communications, which now borders on personal communications service (PCS), involves handheld transceivers. In this application, satellites in low earth orbit (LEO) (altitudes ≈ 1000 km) and medium earth orbit (altitudes ≈ 10,000 km) are emerging which will lay down multiple spot beams similar to cellular structures in terrestrial cellular systems. Here, however, the cells (spot beams) have motion as the satellite flies over, and the mobile is basically stationary when compared with the rapidly moving spot (cellular) beams.

It is also possible for the spot beams to be programmed to continuously searchlight the terrestrial service areas and remain fixed similar to their terrestrial cellular counterparts. This, of course, requires a more complicated antenna such as a phased array or mechanically slewed antenna and/or altitude control of the satellite bus.

Several companies are proposing LEOs or MEOs for both data and voice delivery. The former will principally be supplied by so-called little LEOs, and the latter will supply both data and voice by big LEOs. Clearly big LEOs will generally be more complicated (and expensive) satellites.

3.1.2 Multiple Low Earth Orbit Satellite Operation

For a single LEO satellite operation, real-time service is not possible since the satellite only remains in view for tens of minutes, where typical orbit periods are in the order of 100 minutes or so. This mode of operation would be more applicable to store-and-forward messaging operations. Satellites operating in this mode are now in operation.

In order to provide continuous coverage, a constellation of satellites must be deployed, and the number depends on the altitude of the satellites and the ground transceiver antenna beam minimum elevation angle required to reduce losses due to shadowing and blockage.

MEOs operating at higher altitudes will have orbital periods in the order of 6 h and remain in view for approximately 2 h. Generally MEOs require fewer satellites in orbit since they can see greater portions of the earth than do LEOs. In addition, antenna elevation angles can be higher.

With LEOs, it is possible to blanket the globe or selected regions (where there is a user market) with spot beams thus simulating terrestrial cellular topology. However, because of the altitudes and the finite antenna aperture sizes possible on spacecraft, the spot beams will encompass large areas (hundreds of miles in diameter) which can be referred to as megacells, as compared to macrocells or microcells of terrestrial cellular systems. Terrestrial cells may range from about 1 to 20 mi in diameter.

Unlike the terrestrial and GSO cellular geometries, the moving cells in LEO operation generally will not straddle high—human density areas continuously and may, during periods of their orbits, witness quiescent activity. Clearly, this occurs over polar and oceanic regions. During the transit over low-density areas, service may be curtailed, or beams may also be turned off.

The spot beams or cells laid down by the LEOs may also interface with the terrestrial cellular structure. For example, a mobile or handheld unit, when in an urban environment where there is cellular service, will probably use the terrestrial service. In remote areas, where cellular service is not present, the user will resort to the satellite-based service. Ubiquitous service will therefore be provided. It may also be possible for the satellite service to alleviate traffic congestion on the terrestrial systems during peak hours. If the standards governing the design of the handset for terrestrial use and satellite are not compatible, dual-mode handsets may be required.

The next step, which may arise in this entire drama of mobile communications, can be providing service to handheld units for satellites in GSO. In principle, this is possible. These systems are being proposed, as mentioned previously. In this application, rather large satellite antennas are required to supply a large EIRP to omnidirectional beam handheld units and large G/T in the satellite to temper the uplink power from the handset.

Figure 3.2 shows a typical satellite coverage of the earth by a beam which blankets the service area. The satellite is located at a height of h above the earth, and α is the minimum elevation angle at which terrestrial transceivers can operate with reliable service and acceptable link margins. Alpha (α) is typically greater than 10°.

Even though it would seem desirable to have the greatest area possible covered, there are problems with this reasoning. One is that the number of users may far exceed the number of channels available from the allocated spectrum. Therefore, it is desirable to invoke frequency reuse similar to that used in terrestrial cellular systems by dividing the coverage into cells via spot beams. The greater the number of cells, the more efficient the frequency reuse. However, at the satellite, the beamlets laid down by the antenna will have to use large apertures, since the spot beam footprint (cell) size is a function of the aperture size and altitude. Also, handoff will have to be made more frequently and rapidly because of the movement of the satellite in low earth orbit.

Figure 3.2
Speckled area is the
service area which
may be partitioned
into a cellular struc-
ture (via spot beams)
in order to invoke fre-
quency reuse.

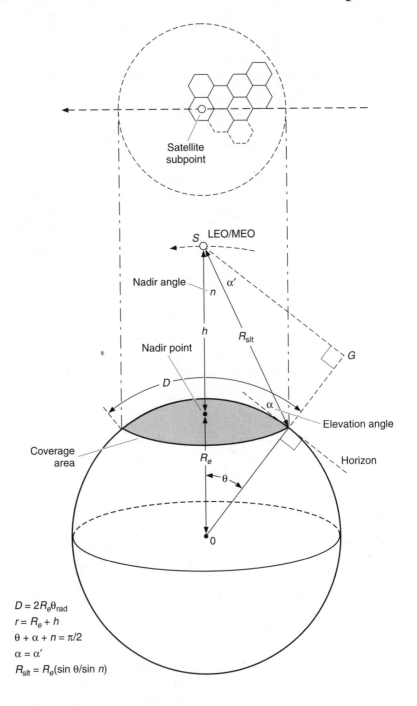

$$D = 2R_e\theta_{rad}$$
$$r = R_e + h$$
$$\theta + \alpha + n = \pi/2$$
$$\alpha = \alpha'$$
$$R_{slt} = R_e(\sin\theta/\sin n)$$

A large antenna lays down a higher power-flux density on the surface of the earth, given by the relationship:

$$PFD = G_sP_s/4\pi R_{slt} \qquad \text{watts/meter squared} \qquad [3.1]$$

The greater PFD means greater power received at the earth terminal since received power is equal to

$$P_{rec} = PFD \times \text{antenna effective area}$$

$$= PFD \times A_e = G_sP_s/4\pi R_{slt}(G_r\lambda^2/4\pi) \qquad \text{watts} \qquad [3.2]$$

where G_s = satellite antenna gain in direction of receiver
P_s = satellite transmitted power
R_{slt} = slant range to ground receiver
G_r = ground receiver antenna gain
A_e = ground antenna effective area

On the downside, the number of cells within a coverage area is limited by the largest antenna which is feasible on the satellite and the launch vehicle shroud size. Here furlable antennas are a solution to realize large antennas.

The cellular structuring in satellite-based mobile communications is to divide the coverage area by multiple spot beams. As in terrestrial cellular systems using frequency division multiple access (FDMA), the users or U per cell or spot beam is therefore

$$U = (1/N)(B_a/B_v) \qquad [3.3]$$

where B_a = allocated spectrum bandwidth
B_v = bandwidth of user channel
N = reuse factor which is dictated by the acceptable cochannel interference (see Glossary for definition of reuse factor)

It is of interest and important to note that the satellite spot beams approximate contiguous cellular clusters as in terrestrial systems, but the performance parameters are not quite the same. First, the satellite cellular signal received at the ground receiver does not manifest the inverse 4th power loss attenuation commonly used in terrestrial cellular systems.[*] Second, in terrestrial cellular systems there is generally no line-of-sight prop-

---

[*] This is an empirical factor based on propagation prediction models generated by Okamura and extended by Hata.

agation, where in satellite applications generally there is line-of-sight propagation, and the signal will have a strong dominant component (plus random components due to multipath). Finally, from an interference point of view, spot beams do not confine their energy to a single spot or cell but wiggle into other cells because of the attendant sidelobes of the satellite beam(s).

Figure 3.3 shows four contiguous footprints (cells) and the satellite antenna spot beams which have generated them. Note that all the beams manifest sidelobe structure and main beam energy which permeate other cells. For example, beam number 1's first sidelobe energy is in cell number 3. Whether cell number 3 can use the same frequency as cell number 1 is dependent on how far down the sidelobe energy is in reference to the bona fide signal in cell number 3. It is also clear that contiguous cells cannot use the same frequency since peripheral cell interference levels are down only about 3 dB (assuming that the contours are touching at the −3 dB contours of the antenna beam).

In other words, practical antenna beams are not idealized sector beams (no sidelobes) and sidelobe energy (interference) will reside in adjacent cells as well as the cells farther out. By weighting the antenna illumination function incident on the aperture of the satellite, one can reduce the sidelobes. Simultaneously, however, there is a concomitant distension of the main lobe and reduced on-axis gain. Therefore, some kind of tradeoff is required.

3.2 Orbit Geometric Relationships

The coverage or spherical area of the earth's surface, as shown in Fig. 3.2, within the visibility cone of the angle of 2θ is given by

$$A_s = 2\pi R_e(1 - \cos 2\theta) \qquad [3.4]$$

where R_e = earth's radius (6378 km at equator)
 θ = central angle
and

$$\theta = \arccos (R_e \cos \alpha/(R_e + h)) - \alpha \qquad [3.5]$$

where α = antenna beam elevation angle of earth station; its value determines system performance
 h = altitude of satellite

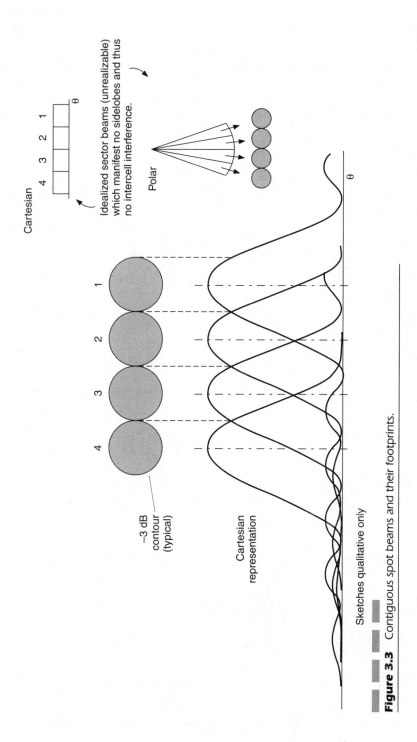

Figure 3.3 Contiguous spot beams and their footprints.

Cartesian

θ

4 3 2 1

Idealized sector beams (unrealizable) which manifest no sidelobes and thus no intercell interference.

Polar

1 2 3 4

−3 dB contour (typical)

Cartesian representation

θ

Sketches qualitative only

53

A plot of relationship as in Eq. [3.5] for parametric values of the elevation angle α is indicated in Fig. 3.4. Note the ordinate is in nautical miles (1 nmi = 1.85 km).

However, as in GSO systems, operation close to zero elevation angle is not used because of the high shadowing and blocking losses and slant penetration through the atmosphere. Low-elevation angles would require higher link margins, which may not be available. It is fortunate that for high-elevation angle operation, the received signal attenuation follows the one over R^2 relationship, and the signal statistics are Rician (dominant component plus random due to multipath). Here margins are nowhere

Figure 3.4
Satellite altitude versus central angle with parametric grazing angle (elevation angle).

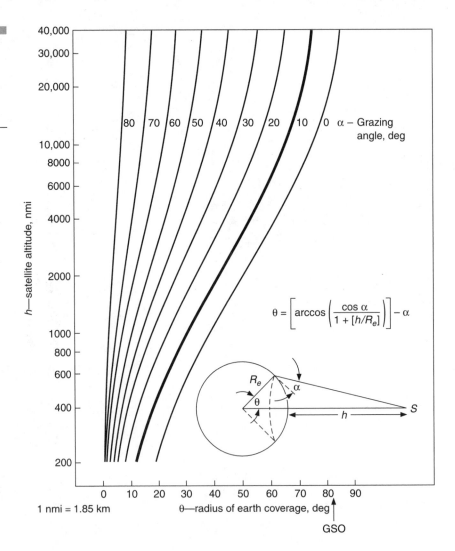

$$\theta = \left[\arccos \left(\frac{\cos \alpha}{1 + [h/R_e]} \right) \right] - \alpha$$

1 nmi = 1.85 km

near values required in terrestrial cellular systems where signal attenuation follows the $1/R^4$ law or worse. In LEO operation, elevation angles better than $20°$ are desirable; if operation is lower than $20°$, an additional margin will be required.

From Fig. 3.2 it is noted that the coverage diameter is given by $D = 2\theta R_e$, where θ is given in radians.

From the preceding information, it is possible to determine the number of satellites required to give full earth coverage at any one time. Clearly, the important terms in this determination are satellite altitude and elevation angles, which in turn determine the satellite beam coverage.

The percentage of the earth's surface in view from the satellite at an altitude for $\alpha = 0°$, the spherical area becomes

$$A_s = 2\pi R_e^2 (h/R_e + h) \tag{3.6}$$

From the law of cosines, the slant range from the satellite to the earth terminal in terms of the central angle θ is

$$R_{\text{slt}}^2 = R_e^2 + (R_e + h)^2 - 2R_e(R_e + h) \cos \theta \tag{3.7}$$

This may also be put in terms of the terrestrial transceiver antenna beam elevation angle

$$\alpha = \arccos [(R_e + h)/R_{\text{slt}}) \sin \theta] \tag{3.8}$$

The distances to the minimum grazing angle point of $10°$ for various altitudes h are shown in Fig. 3.5. Clearly, for greater or smaller grazing angles, the ranges will be smaller or larger, respectively. Note the maximum altitude plotted is below the altitude of the inner Van Allen radiation belt, which commences at about 1500 km altitude. At minimum altitude of 700 km, the air drag can be an influencing factor and can reduce the satellite's orbital life.

In Fig. 3.2 we have formed a triangle with two rays extending from the earth radius and satellite location. We form triangle OGS with a right angle at G. The angles α and α' are identical since they are alternate interior angles formed by parallel lines GS and ray tangential to the earth to form elevation angle α. We therefore have

$$\cos = (h + R_e/R_{\text{slt}}) \sin \theta \tag{3.9}$$

or

$$= \arccos [(r/R_{\text{slt}}) \sin \theta]$$

where $r = h + R_e$, orbital radius.

Figure 3.5
LEO satellite to ter-
restrial transceiver
maximum one-way
distance.

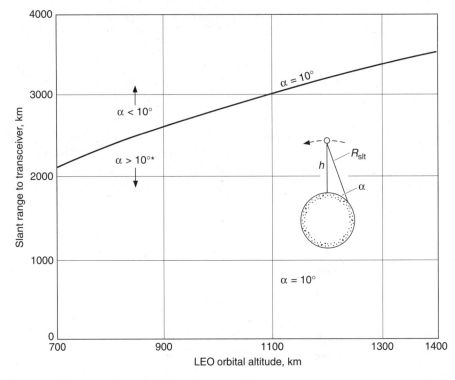

• Terrestrial transceiver at grazing point.
* For example, curve moves downward for grazing angle greater than 10°. At the nadir point, $R_{slt} = h$.

The nadir angle n is given (by law of sines) as

$$\sin n = (\sin \theta) R_e / R_{slt} \qquad \textbf{[3.10]}$$

or

$$n = \arcsin [R_e(\sin \theta / R_{slt})]$$

In Fig. 3.1, the following is true

$$\theta + n + \alpha = \pi/2 \qquad \textbf{[3.11]}$$

Because the ray making up α elevation angle is 90°, we therefore have

$$(\theta + n + \alpha) + \pi/2 = \pi \qquad \text{or } 180° \qquad \textbf{[3.12]}$$

since

$$\sin n = [\sin(\pi - n - \theta) / (R_e + h) \qquad \textbf{[3.13]}$$

we have

$$\sin n = (\cos \alpha) R_e / R_e + h \qquad [3.14]$$

or

$$n = \arcsin [R_e(\cos \alpha / R_e + h)] \qquad [3.15]$$

If $\alpha = 0°$, Eq. [3.15] reduces to

$$n = \arcsin [R_e / R_e + h] \qquad [3.16]$$

or the full angle subtended by the earth is therefore $2n$.

For a satellite in geostationary orbit, the angle $2n$ is fixed at 18.3°. For a satellite in low earth orbit, this angle is a function of satellite altitude. For a minimum practical altitude for a LEO, this angle ($2n$) is in the range of 120°. A plot of the nadir angle and the concomitant central angle as a function of altitude is indicated in Fig. 3.6. This is for the maximum viewing angle or elevation angle equal to 0°.

Another useful curve supplied to the author in a private communication from a colleague* is shown in Fig. 3.7. This curve merges information from Figs. 3.4 and 3.6. In Fig. 3.7, knowing the altitude (usually given) and the desired minimum elevation angle, we can find: (1) earth half angle θ, (2) great circle distance (GCD) of the speckled coverage area, (3) the coverage area, and (4) the nadir angle.

As an illustration, assume a LEO satellite at an altitude of 1500 km and a minimum elevation angle of 10°. Follow the bowed curve at 1500 km to the elevation angle curve intersection and down to the three abscissa axes. We obtain θ = 27.5°, GCD = 2250 nmi (6012 km), and coverage area $S = 28 \times 10^{+6}$ km². The nadir angle is read from the ordinate axis at the 1500 km level projected horizontally to the vertical axis. One reads 54°.

It is clear that the one-way time delay between the satellite and transceiver is

$$\tau = (R_{slt1} + R_{slt2})/c \qquad [3.17]$$

where c = velocity of EM wave = 3×10^8 m/sec
         R_{slt1}, R_{slt2} = uplink and downlink slant ranges

The time of transmission to and from the satellite and ground transceiver for LEOs is significantly less than for satellites in GSO. One-way

---

* Private communication from R. Z. Zaputowycz, Bell Atlantic NYNEX Mobile.

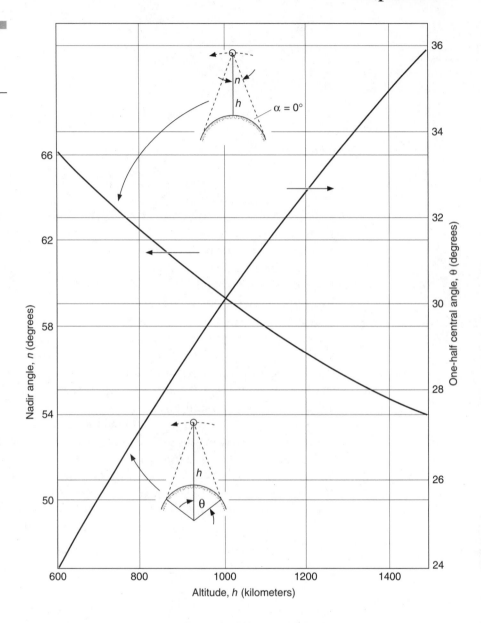

delays for GSO satellites are between 240 and 270 ms. For a satellite in LEO at an altitude of say 1000 km (and $\alpha = 10°$), the one-way propagation delay is in the order of 19 ms or less. For satellites in MEO, the delay is about 90 ms. The delays are not significant enough to cause echo problems similar to those encountered in satellites in GSO. Echoes can be suppressed. The problem is voice delay making two-way conversation difficult in 500-ms round trip delay situations.

Figure 3.7
Coverage parameters.

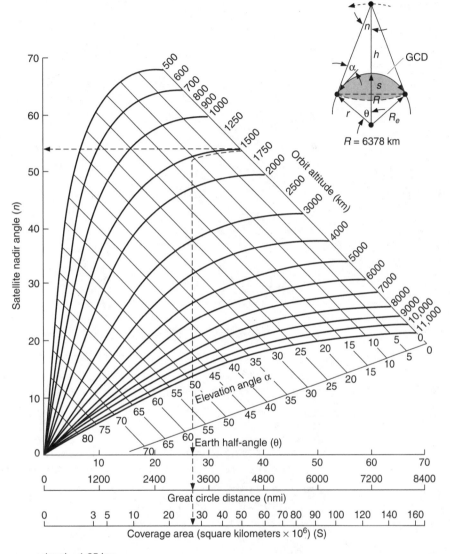

• 1 nmi = 1.85 km

The minimum grazing angle of 10° is frequently used to establish the criterion to achieve adequate performance. Having established the minimum elevation angle acceptable to give adequate performance (with a reasonable link margin), we can determine the maximum time a satellite is in view of the terrestrial transceivers. If we accept the minimum angle to be 10° (for example), we can find the maximum viewing time as a function of satellite altitude. This is plotted in Fig. 3.8.

Figure 3.8
Maximum viewing
angle for a satellite
overflight of terres-
trial transceiver.

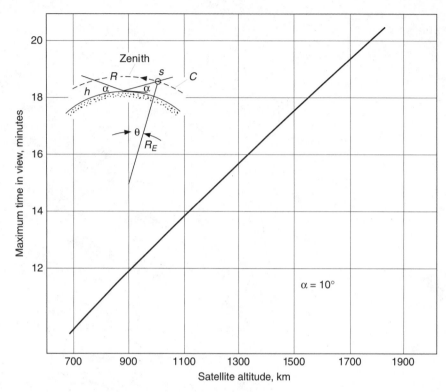

- Terrestrial transceiver antenna beam elevation angle equals 10° (within coverage area).
- For higher viewing angles, the curve is translated downward.

Note that the maximum viewing time is achieved if the satellite over-flies the transceiver, or passage is over the zenith position. For all other flight paths, the viewing times will be less.

The circumference of the orbit is

$$C = 2\pi(R_e + h) \qquad [3.18]$$

The part of the orbit which is in performance view for $\alpha \geq 10$ is given by

$$R = (R_e + h)2\theta \qquad [3.19]$$

The part of the orbital period T in which the satellite is in an acceptable view angle is therefore

$$(R/C)T = \frac{(R_e + h)2\theta}{\pi(R_e + h)^2} \times T_p = \tau$$

$$\tau = \frac{2\theta}{\pi(R_e + h)} \times T_p \qquad [3.20]$$

It has been shown previously

$$\theta = \arccos\left[(R_e/R_e + h)\cos\alpha_{max}\right] - \alpha_{max} \qquad [3.21]$$

where $\alpha_{max} = 10°$.

It is obvious that in urbanized areas where blockage may present a severe problem, viewing angles may be less, and time in view is correspondingly less.

For a mobile satellite in geostationary orbit, the altitude above the equator (the nadir point) is 35,780 km. An earth station located at the earth's limb has a distance of 41,756 km. The one-way difference in signal loss between the two points is given by expression [3.22].

The free space loss is

$$L_{FSL} = 10 \log [4\pi R_{slt}/\lambda]^2 \qquad [3.22]$$

The differential loss between the satellite subpoint or nadir point and the limb is therefore

$$\Delta L = 10 \log \left[\frac{4\pi R_{slt}}{\lambda} \bigg/ \frac{4\pi R_n}{\lambda} \right]^2$$

$$= 10 \log [R_{slt}/R_n]^2 \qquad [3.23]$$

Plugging in the values given in Eq. [3.23] for the two ranges (to limb and to nadir point), we obtain the differential loss

$$L = 1.3 \text{ dB} \qquad [3.23']$$

For mobile satellites in LEO or MEO, the differential losses between the satellite subpoint and the limit of the coverage area are a function of the satellite altitude and elevation angles. The losses with elevation angle being parametric are shown in Fig. 3.9. For a typical LEO altitude of 1 km and an elevation angle of 10°, the differential loss is about 9 dB. This stems from the relationships shown at the bottom of the graph.

For a simple satellite antenna, the beam width is rather extreme to cover the full area. For example, for an altitude of 1000 km, the satellite angle (twice the nadir angle) subtends the earth out to the horizon by about 120° (see Fig. 3.6). For a single beam antenna, the beam is 120° wide and must be shaped to provide constant power-flux density coverage

Figure 3.9
Differential slant range free space losses as a function of satellite altitude. Elevation angle is parametric.

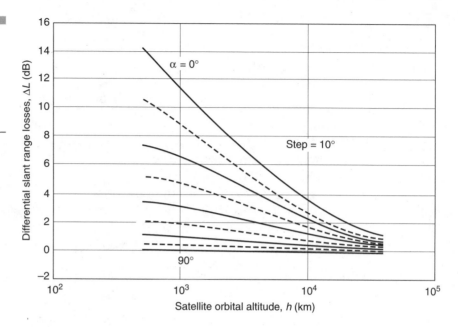

• Parametric elevation angle α reflects wide differences in slant range and free space losses to users near the nadir point and the periphery of the coverage area.

• An octave change in range changes the loss by 6 dB. Some of this loss is compensated for by adaption of shaped pattern beams on the satellite and the terrestrial transceivers, or distance-dependent (nadir point epoch) spot beam gain for systems using spot beams. That is, increasing the gain of the spot beams as they deploy farther from the nadir point helps to maintain a nearly isoflux power density across the coverage area.

between the nadir point and the horizon. That is, the beam shape is concave with 9 dB less gain at nadir than at the horizon.

In an operational sense, when the coverage area is cellularized, the cells located at the periphery of the coverage bounds will receive 9 dB less power ($\alpha = 10°$) than for cells at the satellite subpoint. It therefore suggests that both the satellite antenna beams (spots) and the earth station antenna beam be shaped to prevent wide variations in signal levels at the receiver.

Spaceborne antennas (arrays in particular) can be tailored to produce variable gain spot beams. Ground transceiver antennas may use bifilar and quadrifilar helices providing shaped beams as well as omnidirectional coverage with circular polarization. LEO and MEO satellites will use circular polarized satellite antennas in their communication channels to facilitate receiver antenna pointing and reduce propagation anomalies. Quarter wave stub antennas on the handsets may be used, but these are linearly polarized and would only accept one-half of the power from a downlinked circularly polarized signal. In uplinking, some power would also be lost at the satellite receiver antenna.

For the single satellite orbit shown in Fig. 3.1, the cellular sizes in the cluster are a function of the satellite altitude and the antenna size. Real-time communication is possible between cells in the same coverage (see Fig. 3.10), but only for short periods of time. For service to different service areas, store-and-forward or mail box operation is necessary with an intervening time interval. This interval may be in hours depending on whether the receiver is in the same terrestrial swath or another region of the globe.

Because of the earth's rotation, the satellite will not pass over the same ground track as the previous orbit. The earth rotates through one revolution in its sidereal period T_e in 24 h. The satellite period is T_s. Therefore, the earth's rotation in one orbital period is

$$\Delta\theta = -2\pi(T_e/T_s) = -2\pi(T_e/T_s) \qquad \text{radians/orbit} \qquad [3.24]$$

For example, $T_s = 1.5$ h. Therefore

$$\Delta\theta = -2\pi(1.5/24) = -0.393 \text{ rad/hr} = -22.5 \text{ degrees}^*$$

There will be an additional translation of the nodal crossing (where the satellite crosses the equator in the northerly direction) due to the nodal regression resulting from the earth's perturbations (oblateness). The amount of regression $(i < 90°)$ is a function of inclination and satellite altitude. Nodal regression does not exist for inclinations equal to $90°$.

---

* At the equator, a point moves eastward at about 1660 km/h. At increasing latitudes, the locations move more slowly.

Figure 3.10
Two transmission
modes for satellites
in LEO/MEO.

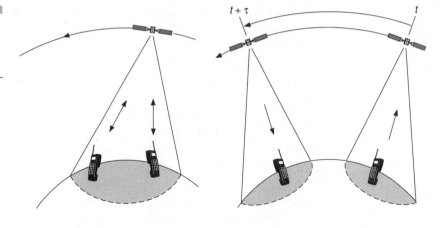

Real-time communications   Store-and-forward communications

3.3 Global Cellular Constellation Design

To meet the performance requirements for a full global cellular coverage, a constellation of satellites is required since a single satellite system will not provide full coverage nor be available on a continuous basis. Clearly, for economic reasons, the fewer the number of satellites in the constellation, the better. Two other factors come into play: the altitude of the satellites in the constellation and the minimum usable elevation angle of the antenna beam of the terrestrial transceiver (handsets). Both the altitude of operation and the elevation angle have impact on the number of satellites required. Intuitively, one can see that the higher the orbital altitude (within limits), the greater the viewing area and the fewer the number of satellites needed to cover the earth on a continuous basis. Similarly, the higher the elevation angle to reduce the shadowing and blocking losses, the greater the number of satellites required. More satellites are packed into the orbits since their viewing angles have been restricted to satisfy minimum elevation angles.

Two schools of thought have evolved in constellation design for low earth orbit and medium earth orbit. One is deploying satellites in multiple polar orbits ($i = 90°$) or near polar orbit. Research in this area has been performed by several investigators (Beste, Adams, and Rider). The other approach has considered satellites in several planes but in inclined orbits. These studies have been completed by Ballard and Walker.

The constellation designs have been most tractable with satellites in circular orbits to avoid orbit dynamic problems which accompany elliptical orbits. Circular orbits do not manifest the regression of the line of apsides (rotation of orbital plane), as would occur for elliptical orbits if the inclination were not 63.4° or 116.6°. However, circular orbits do demonstrate nodal regression and the amount depends on the inclination and altitude of the satellite(s).

3.4 Polar Orbit Constellations

In polar orbiting satellites, the satellites pass over the poles or have inclinations* close to 90°. The geometry for one satellite is depicted in Fig. 3.11. The swath width is the coverage area for a satellite in the orbital plane. The upper bounds on the coverage area (dictated by the elevation angle of 10 and the altitude). The swath width, S, is shown in Fig. 3.12 as a function of elevation angle with altitude being parametric. The field of view extends out to the horizon as also shown on the figure. Here the elevation angle is equal to 0°. As indicated previously, there is no nodal regression, but westerly movement will occur due to the earth's rotation, which rotates counterclockwise when viewed from the north pole.

In polar orbit constellations, a number of satellites are launched into circular orbits with an equal number in each plane. The orbits are of sufficient altitude to mitigate the air drag effects and low enough to avoid the exposure to the inner Van Allen radiation belt.

A classical bit of work in the design of polar orbit constellations is based on a paper by Adams and Rider. Motorola, in the design of their big LEO constellation (Iridium), used the analysis of Adams and Rider.

The Iridium system will use three antenna arrays (see Fig. 3.13) with each producing 16 cellular spot beams, and the beams are juxtaposed within the coverage area. There will therefore be 48 spot beam footprints within the coverage area. The diameter of these footprints is rather large and in the order of hundreds of kilometers. It is safe to call the cells they produce megacells. The 48 spot beams will travel very rapidly along the surface of the earth, and the mobile users will appear to be basically stationary. The mobiles will therefore see different spot beam footprints (or

---

* It is important that all satellites have the same inclination so that they nodally regress at the same rate. Lines of nodal regression (move westward) for $0 < i < 90°$ (prograde orbit).

Figure 3.11
Polar orbiting low
earth orbit satellite.

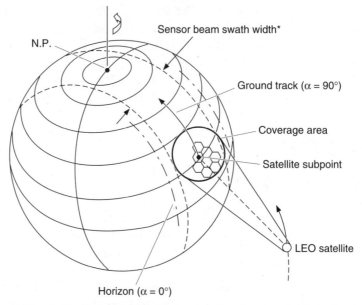

N.P.

Sensor beam swath width*

Ground track ($\alpha = 90°$)

Coverage area

Satellite subpoint

LEO satellite

Horizon ($\alpha = 0°$)

* Due to earth's rotation, the swath shifts westward and is a function of altitude. Swath width may or may not be equal to field of view (FOV), or visual width. There is no nodal regression for $i = 90°$.

The footprint size is designated by the minimum elevation angle desired at the coverage periphery. Clearly, all spot beams (cells) within this coverage will provide equal or greater elevation angles to all users.

cells). The handoff from one cell to the other will be very rapid, but the cell in which the user or mobile will be next will be known a priori to the satellite operator.

Some of the salient parameters of the Iridium satellite constellation are as follows:

Number of satellites: **66** (originally was 77)

Number of circular orbital planes: **6**

Number of satellites per plane: **11**

Inclination: **86.4°** (originally was 90°)

Altitude of orbits: **780 km**

Minimum elevation angle at edge of coverage: **8.2°**

A typical spot beam coverage of the earth for satellites in polar orbits is shown in Fig. 3.14. This was taken from Motorola's initial application to the FCC for their Iridium system.

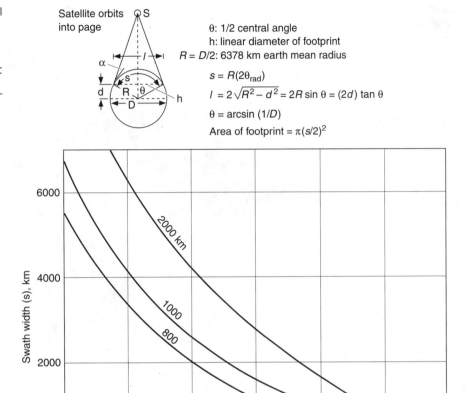

Figure 3.12

Swath width as a function of elevation angle with parametric altitude.

Note that for a polar constellation there is a greater density of spot beam footprints near the poles, and ground transceivers can see more than one satellite, whereas at lower latitudes, generally only one will be visible.

The global coverage for a 66-satellite constellation is shown in Mercator projection in Fig. 3.15. Here the inclination is 86° and is reflected by the bounds of the plots in the latitudinal directions. Each coverage plot in Fig. 3.15 is partitioned into 48 cells or spot beams.

Each of the 66 coverage beams generates 48 spot beams for a total of 48 × 66 = 3168 global spot beams. Only 2150 of these spot beams are active due to the exclusion of some of the overlapping beams which

Figure 3.13
Sketch of the Iridium
satellite design.

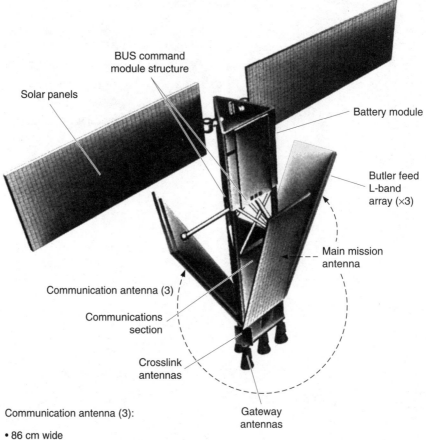

BUS command
module structure

Solar panels

Battery module

Butler feed
L-band
array (×3)

Main mission
antenna

Communication antenna (3)

Communications
section

Crosslink
antennas

Gateway
antennas

Communication antenna (3):

• 86 cm wide
• 186 cm high
• 4 cm thick
• 106 radiating elements
• 16 beams per antenna
• 48 beams juxtaposed

occur near the poles (these are turned off). Each cell or spot beam has a diameter of 600 km (370 mi). The overall footprint diameter encompassing the cells is about 4700 km.

As indicated previously, the cells move very rapidly over the surface of the earth, and the mobiles, for all practical purposes, are stationary. The ground track velocity (see Fig. 3.11) of a cell cluster within a coverage area is given by

$$v = 2\pi(R_e + h)/T_s \qquad [3.25]$$

Figure 3.14
Iridium multiple spot beams (cells) blanketing the earth's surface (original design using 77 satellites).

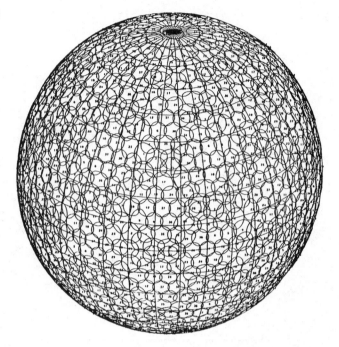

- 77 satellites
- i=90°

where R_e = earth's radius (6378 km)
   h = orbit altitude (780 km)
   T_s = orbital period ($\approx$100 min)

For circular orbits the period is

$$T = 2\pi\sqrt{r/u} \qquad\qquad [3.26]$$

where r = orbital radius $(R_e + h)$
   $u = GM$, gravitational parameter (for earth this value is 398,601.2 km^3/s^2)
   $\sqrt{u} = 631.35$ km$^{3/2}$/s
   $R_e = 6378$ km (equatorial radius)

The cluster therefore moves at a velocity of about 7.5 km/s (16,740 mi/h).*

---

* If the mobile average velocity is 50 mi/h, we can safely say that the mobile is essentially stationary compared to the sweeping beams.

Figure 3.15
Mercator plot of a
66-satellite global
coverage.

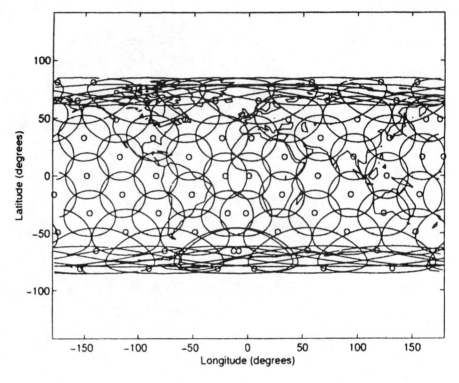

- For IRIDIUM, each coverage area supports 48 spot beams.

The polar constellation used by the Iridium system is shown in Fig. 3.16 [Benedicto et al. and Larson and Wertz]. The large figure shows a polar view with the angular spacing between adjacent planes. The corotating planes are separated by 31.6°. There is an area between the first and last planes where the satellites are counterrotating. This has been referred to as the counterrotating seam. The separation of the planes which form this seam is 22°.

The satellites are staggered in adjacent orbits so that the center of the antenna beam is separated by π/N, where N is the number of satellites in the plane and, in this case, 11 satellites per plane. The plane separation for corotating satellite planes is $= \theta + \varepsilon$, where θ is the central angle (see Fig. 3.1) and ε is the one-half width of the ground swath, as shown in Fig. 3.16. The plane separation for the adjacent counterrotating satellites is equal to twice the ground swath as $\Delta_2 = 2\varepsilon$, or $\Delta_2 = 22°$ in this case. The coverage area packing in proximity to the equator is shown in Figs. 3.17 and 3.18.

Figure 3.16
Near polar constella-
tion used by Iridium.

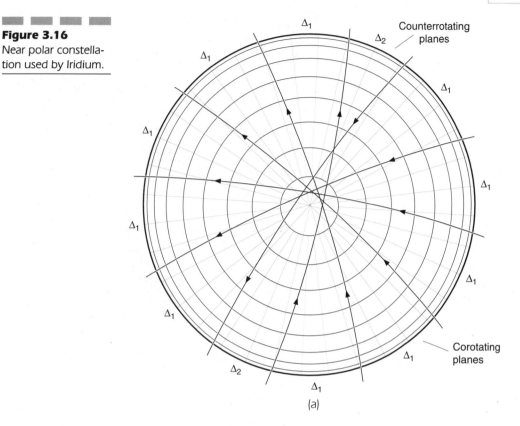

(a)

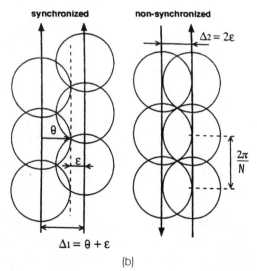

(b)

Figure 3.17
A closer look at the
equatorial coverage
for Iridium.

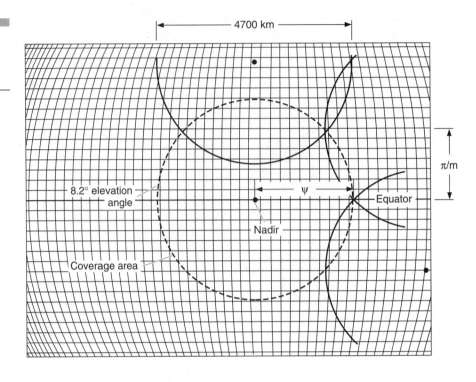

Each circular contour supports 48 spot beams (cells) and the cells in proximity to the coverage boundary will have an elevation angle of 8.2° relative to its parent satellite. As indicated previously, the angular separation between the center of coverage areas in adjacent planes is π/N. The plane separation is $\theta + \varepsilon$.

Motorola indicates that the width of the coverage area is about 4700 km. The width of the nadir spot beam is about 600 km. As the 48 spot beams illuminating the earth deploy farther from the nadir point, the footprints distend since the spot beam ground intercept lands on a curved earth. Actually, the beam contours at the periphery of the coverage area would open up tulip fashion if the coverage were to extend out to the horizon. See the section on cellular structuring in Sec. 3.6.

Adams and Rider and Benedicto et al. have shown that for a polar constellation and full global coverage, one obtains the following orbital parameters:

Figure 3.18
Optimum global coverage with optimum orbital spacing and intraorbit satellite phasing (separation).

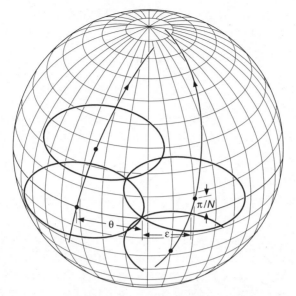

• Orbital planes are separated by θ + ε.
  The separation at the seam is 2ε.

Number of satellites per plane:

$$N \approx (2/\sqrt{3})(\pi/\theta) \qquad [3.27]$$

Number of planes:

$$P \approx (2/3)(\pi/\theta) \qquad [3.28]$$

Number of satellites in the constellation:

$$T_c \approx (4\sqrt{3}/9)(\pi/\theta)^2 \qquad [3.29]$$

where θ is the central angle (see Fig. 3.2).

The total number of satellites is dependent on the central angle θ. The *larger* the central angle to reduce the number of satellites required, the *lower* the elevation angle will be. There is, therefore, a tradeoff between the minimum number of satellites and the smallest elevation angle that will achieve adequate performance. As indicated previously, the smaller elevation angle will offer greater losses due to shadowing and blockage, necessitating a larger fade margin in the link equation. Power in a satellite (and terrestrial transceivers) is at a premium.

The central angle θ and the angle half width of the ground swath,* ε (see Fig. 3.15) are determined by

$$\cos \theta = \cos \varepsilon \cos (\pi/N)$$

$$\text{or } \varepsilon = \arccos \left[\cos \theta / \cos (\pi/N) \right] \qquad \text{[3.30]}$$

where N is the number of satellites in an orbital plane.
The central angle for the Iridium system is 20°.

3.5 Inclined Orbit Constellations

The other scheme to determine the number of satellites needed to cover the globe is to use satellites in inclined orbit. Polar orbits have a greater concentration of satellites near the poles, whereas an inclined orbit approach tends to give a more uniform distribution globally. This design approach is used by Loral in their Globalstar LEO system and TRW in their Odyssey MEO system. This constellation design was extensively studied by Walker and used by Loral and TRW.

The constellation consists of a total of T satellites with S satellites evenly distributed among P planes. All planes have the same direct inclination ($i < 90°$) and altitude, and all undergo the same nodal regression. The ascending nodes of the P planes are evenly distributed around the equator at intervals of $360°/P$. In each plane, the satellites are uniformly distributed at intervals of $360°/S$. There is also a phasing requirement of satellites in adjacent planes given by $360°F/T$, where F is the phasing parameter. The above relationships are depicted in Fig. 3.19. A Walker constellation is usually designated by the symbolism

$$i : T/P/F \qquad \text{[3.31]}$$

where i = inclination of circular orbits
T = total number of satellites
P = total number of orbital planes with the same inclination
F = phasing factor, $0 \leq F \leq P - 1$

The number of satellites S is evenly deployed or evenly phased. Any disparity in phasing can be corrected by using a small amount of onboard

---

* In their paper, Adams and Rider refer to this as one-half street width.

■■■ ■■■ ■■■ ■■■

Figure 3.19
Walker's notation
and constellation
definition.

Constellations denoted by T/P/F

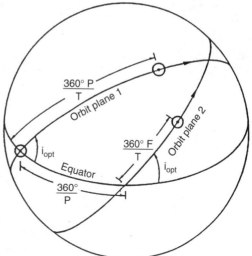

T = Total number of satellites
P = Number of orbit planes
F = Phasing parameter

T satellites are equally divided
among P planes.

P planes are equally spaced in Ω.

Satellites within a plane are equally
spaced in argument of latitude.

$F \times \frac{360°}{T}$ = phasing difference between
satellites in adjacent planes. See
sketch.

i_{OPT} = Optimized inclination that is common to all planes.

R_{max} or d_{max} = Coverage circle radius required by a given constellation. For a given
elevation angle, R_{max} defines the minimum common altitude for all the satellites in
the constellation.

D_{min} = Minimum earth central angle between any two satellites in the constellation.

fuel to raise* or lower the altitude of the satellite. After intrasatellite orbit
separation is achieved, they are returned to their original orbit. Intraorbit
phasing requires considerably less propellant than if an orbital plane
inclination change were required. At launch, a single launch vehicle used
to launch several satellites will put them into a single plane.

An example of a Walker constellation is represented in Fig. 3.20. As indicated, the inclination is 65° for five orbital planes with a phasing factor
equal to 1. The plane difference between satellites in adjacent planes is
therefore $(F \times 360°)/T = (1 \times 360/15) = 24°$. The number of satellites per plane
is $S = T/P = 3$, with its intraorbit spacing of 120°. The node spacing is 72 (5
planes).

In the Loral Globalstar constellation, a total of 48 satellites are used in
eight orbital planes with six satellites per plane. The satellites are in circular orbits with an inclination of 52° at an altitude of 1414 km. The phasing
parameter is 1. Using the Walker designation, we therefore have $i:T/P/F =$

---

* By Kepler's third law of planetary motion, raising the satellite altitude will reduce its
velocity and will drift backward with respect to a satellite in a lower orbit. This will increase
the distance from the satellite ahead of it in a lower orbit.

Figure 3.20
A 15/5/1 Walker con-
stellation at 65°
inclination.

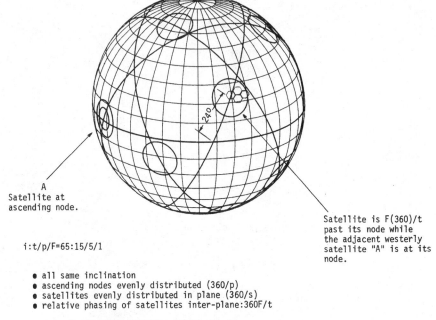

```
                    A
              Satellite at
              ascending node.

                                              Satellite is F(360)/t
                                              past its node while
                                              the adjacent westerly
                                              satellite "A" is at its
                                              node.

              i:t/p/F=65:15/5/1

                ● all same inclination
                ● ascending nodes evenly distributed (360/p)
                ● satellites evenly distributed in plane (360/s)
                ● relative phasing of satellites inter-plane:360F/t

                      F: interplane phasing factor
                      p: number of planes
                      t: total number of satellites
                      S: satellites per plane: t/p, evenly spaced
```

52:48/8/1. The Loral Walker constellation is shown in Fig. 3.21, and its global coverage is shown in Fig. 3.22. Note that even though the inclination of the orbit is 52° and the maximum latitude excursion of the satellite is ±52°, the coverage exceeds these latitudes going up to ±75°. The users in each of the contours shown will have elevation angles greater than 15°, and the contours are blanketed by six spot beams.

A MEO system using the Walker constellation is TRW's Odyssey system. The altitude of operation is at 10,370 km, which is between the two Van Allen radiation belts. A total of 12 satellites are deployed into three orbital planes. The orbits are circular with inclination of 55°. Its Walker designation is 55:12/3/1.

The Odyssey constellation is indicated in Fig. 3.23. The satellite has global coverage areas indicated in Fig. 3.24. Even though the satellite will be visible for about 2 h (in a 6-hour orbit), service will be for about 1 h when the elevation angles are better than 30°.

Each coverage area is partitioned into 37 spot beams as shown in Fig. 3.24. The footprints are shown distorted because they are impinging on a curved earth.

▬▬ ▬▬ ▬▬ ▬▬

Figure 3.21
Loral/Qualcomm
Globalstar big LEO
Walker constellation.

```
i:t,p,F = 52°:48/8/1
h: 1414 km
T: 113 minutes
```

Unlike in the Iridium and Globalstar systems where coverage areas are always nadir pointing, the satellite subpoint will not be generally centered on the coverage area. But, the composite beam will be satellite-attitude controlled (antennas fixed to the satellite bus) to remain on the designated service area, even though the satellite is in its orbital motion. Thus, the user and populated areas will remain in focus for a longer period of time. This has been referred to by TRW as the directed coverage. This concept is depicted in Fig. 3.25.

In September 1994, Inmarsat announced the formation of a new company to build a constellation of satellites operating at an altitude of about 10,300 km. It was initially dubbed Project 21, but now referred to as Intermediate Circular Orbit (ICO) Global Communications. The constellation will consist of 10 satellites (plus two spares) in two or three circular orbits with inclination of 45° and will provide global and continuous telephone and data service. Service is expected to commence at the start of the next century.

Figure 3.22
Global coverage of
Globalstar's 48-
satellite constellation.

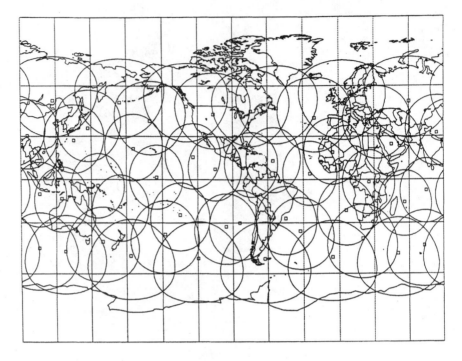

- Squares reflect sub-satellite locations.
- Contours, minimum elevation angles of 15°.
- Inclination of 52°, but coverage to ±70° latitude.
- Two satellite coverage in temperate zones.
- Each coverage area will be blanketed by 16 spot
  beams (16 receive beams on the uplink at about
  1.6 GHz, and 16 on the downlink at about 2.5 GHz).

The global coverage by the 10 satellites is depicted in Fig. 3.26. Each contour represents the edge of coverage of 10°. Other sources indicate that the elevation coverage may be 20°. If this is the case, the contours will tighten up. Each footprint is partitioned into spot beams (163) representing cells. Additional parameters for this system are given in Chap. 7.

3.6 Cellular Structuring

As alluded to previously, the coverage area (which can cover quite a bit of territory) is partitioned into cells via spot beams using a multibeam antenna. This can therefore take advantage of frequency reuse. The coverage area broken down into cells is shown in Fig. 3.27. A mobile user in a cell is essentially fixed as the spot beam footprints swath the earth. Therefore,

Figure 3.23
TRW's Odyssey MEO
satellite constellation.

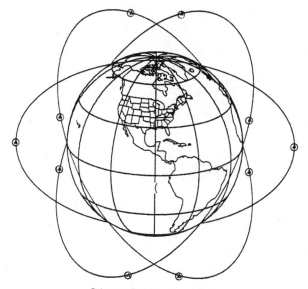

Odyssey Satellite Constellation

- Number of satellites: 12
- Number of planes: 3
- Altitude (circular): 10,370 km*
- Inclination: 55°
- T=6 hours

* Between the two Van Allen radiation belts.

the satellite operator knows a priori into which cell there will be handover. For example, in the sketch shown previously, mobile A moves into cells B and C and eventually into D. In terrestrial systems, the cells are fixed, and the direction of the mobile movement is unknown and may move in one of four directions or not at all.

The handoff is more network challenging for satellite-based cellular systems than for terrestrial systems. The satellite track moves very swiftly, and the mobile is quickly transferred from one cell (spot beam) to another.

As was shown in Eq. [3.25], the ground track velocity for a satellite at an altitude of about 1000 km will be in the order of 7 to 8 km/s or about 17,000 mi/h.

As indicated in previous sections, the coverage area covered by a satellite for central angle θ is given by Eq. [3.4] which is repeated here.

$$A_s = 2\pi R_e(1 - \cos 2\theta) \qquad [3.4]$$

where θ = central angle
R_e = earth's radius

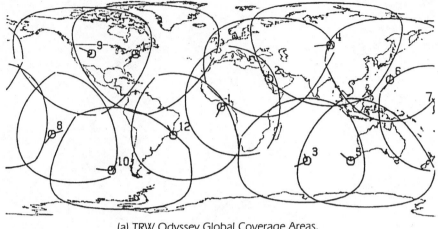

(a) TRW Odyssey Global Coverage Areas.

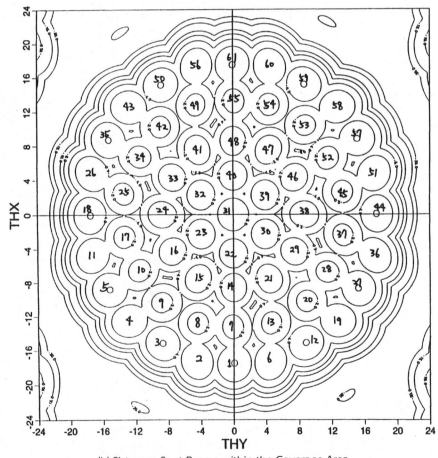

(b) Sixty-one Spot Beams within the Coverage Area.

Figure 3.25
TRW's concept of directed coverage of the service area.

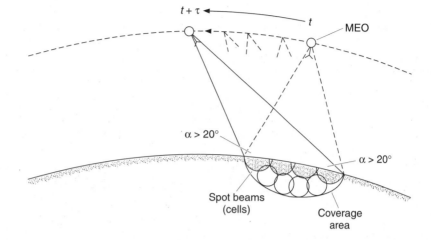

- The satellite bus is attitude controlled [communication antenna (L/S GHz) is fixed to bus] so that the assigned service area is continuously searchlighted by the spot beams as the satellite is in orbital flight.

- The time of searchlighting is limited to about 1 h to realize high-elevation angles even though the satellite will be in view for about 2 h (MEO has a 6-h orbit).

- This mode of operation reduces the requirement for frequent handover from beam to beam or satellite to satellite. A call may even be complete before handover is required. This is coupled with the fact that MEOs pass overhead more slowly than LEOs. This is in contrast to systems in which the spot beams are continuously nadir pointing.

Figure 3.26
Global coverage of ICO's 10-satellite constellation.

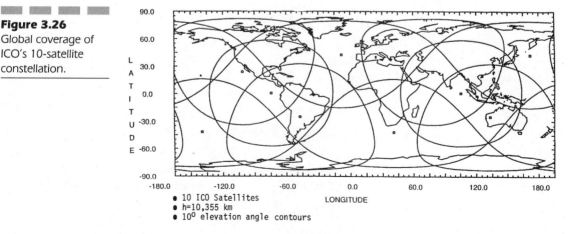

- 10 ICO Satellites
- h=10,355 km
- 10° elevation angle contours

Figure 3.27
Hexagonal tessella-
tion of the dynamical
coverage area.

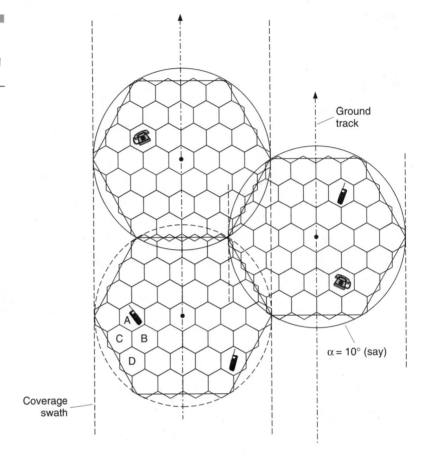

Note: The coverage area cellularization also moderates the power requirements
(satellite and handset), in addition to permitting frequency reuse.

The value of θ is chosen so that at the boundary of this area the elevation angle of the user's transceiver antenna beam will be adequate to give
reliable performance. The minimum angle of coverage can be found
from the expression [3.5],

$$\cos (\theta + \alpha) = (R_e \cos \alpha)/R_e + h \qquad [3.2]$$

where α = minimum acceptable elevation angle.

Each coverage area, as shown in Fig. 3.27, is partitioned into tessellated
cells. In reality, these cells represent contiguous spot beams. However,
symmetrical spot beams (pencil beams) which typically illuminate the
earth will not produce footprints which are circular but distorted

Figure 3.28
Footprint distension
as the spot beams
generated are
deployed away from
the satellite subpoint.

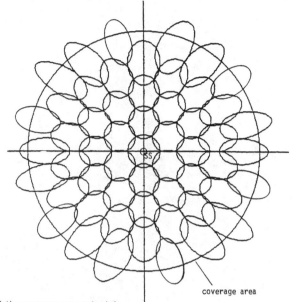

coverage area

Note: If the coverage area extended
out to the horizon ($\alpha=0^{o}$), the
outer footprintsextremies would
fall off the face of the earth.
The footprints would open up.

because of the earth's curvature and slant angle. Only with the nadir
beam (at satellite subpoint) will the footprint be circular. For the beams
which progressively move away from the nadir point, the footprints
become increasingly distorted. An example of this phenomenon is
shown in Fig. 3.28. This is not for the cellular structure shown in Fig. 3.27.
Note that the footprints in proximity to the satellite subpoint are circular but become more egg shaped as they move toward the coverage area
boundary.

In Fig. 3.29 the footprint width in the radial direction away from the
satellite subpoint is:

$$
\begin{aligned}
w &= S_2 - S_1 \\
&= R_c\theta_2 - R_c\theta_1 \\
&= R_c(\theta_2)(\pi/180) - R_c(\theta_1)(\pi/180) \\
&= R_c(\pi/180)(\theta_2 - \theta_1) \\
&= 111.32(\theta_2 - \theta_1)
\end{aligned}
\qquad [3.33]
$$

where the central angles $\theta_{1,2}$ are in degrees.

Figure 3.29
Satellite spot beam
distension as the spot
beam is deployed
from the nadir point
(satellite subpoint).

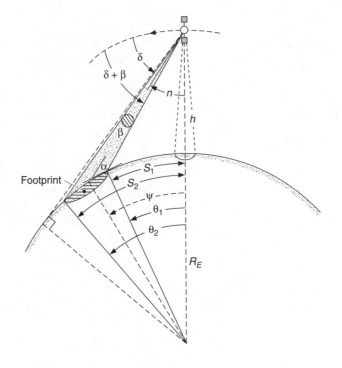

β: Spot beam width in elevation
n: Nadir angle
δ: Depression angle

• Spot beams may be axis-symmetric (pencil beams),
  but they may also be shaped.

The central angle is given as:

$$\theta_{1,2} = \arcsin\left[((R_c + h)/R_c)\cos\phi\right] - 90 + \phi \tag{3.34}$$

where ϕ is the depression angle (either δ or $\delta + \beta$).

If we plug Eq. [3.35] into Eq. [3.34] we obtain the width of the footprint smear (because of the curvature of the earth) of the spot beam across the earth in the radial direction.

$$w = (\pi R_c/180)\langle\arcsin\left[((R_c + h)/R_c)\cos\delta\right] - \cancel{90} + \cancel{\delta}$$
$$- \arcsin\left[((R_c + h)/R)\cos(\delta + \beta)\right] + \cancel{90} - \cancel{\delta} - \beta\rangle \tag{3.35}$$

The footprint smear on the surface of the earth is therefore a function of the spot elevation beam width β, beam depression angle δ, and satellite altitude.

It is of interest to note that if the footprint bounds went beyond the horizon or the limb of the earth, the footprint would open up. This was alluded to previously as the so-called tulip effect of the footprint as it falls off the edge of the earth.

The slant range losses increase as the footprint moves farther from the satellite subpoint. This differential loss can be found from Eq. [3.10] and from Fig. 3.9. For example, for a cell or spot beam at an elevation angle of 30° and satellite altitude of 1000 km, the differential loss is about 5 dB. One can compensate for this loss by increasing the spot beam antenna gain as the beam moves farther out. This would complicate the onboard antenna design. Nevertheless, this compensation is being done in the Iridium LEO system and Globalstar LEO system.

We can make further interesting observations if we refer to Fig. 3.28. From the law of sines, we can find the slant distance at which the beam axis intercepts the earth.

$$R_{slt}/\sin \psi = R_e/\sin n$$

$$R_{slt} = R_e(\sin \psi/\sin n) \qquad [3.36]$$

If we know the elevation angle α, we can find the nadir angle of the spot beam.

$$\sin n = \cos \alpha \sin \Omega \qquad [3.37]$$

where Ω equals the nadir angle out to the horizon ($n < \Omega$) and $n + \psi + \alpha = 90$.

3.7 Discussion

With the advent of LEO/MEO constellations, personal communications networks provide ubiquitous global personal service communications (PSC). They are an ideal medium for providing personal communications to remote areas where terrestrial cellular service is not accessible, or to complement existing cellular systems where traffic density is high.

Satellite-based cellular communication is not unlike its terrestrial cellular counterpart. Cellular structuring is made possible by the development of satellite antennas which can concentrate power into cellularlike footprints. This, as in terrestrial applications, makes it possible to increase the system capacity by the use of frequency reuse and furthermore makes communications possible to handheld transceivers. The cells are much larger (megacells), since their size is limited by the spaceborne antenna size. The larger cells also limit the number of cells in the coverage area and the reuse factor.

The problem of handover is also evident in satellite wireless service. Here, however, in contrast to terrestrial systems, the cells are in motion, and the user is relatively fixed compared to the rapid movement of the spot beams (cells). However, there is solace in the fact that it is known a priori into which cell the user will be since the spot beams scan the terrain in a known direction.

LEOs and MEOs, because of their operation at a lower altitude than the GSO satellites, provide lower attenuation to the uplink and downlink signals in addition to lower signal delays. Even though the signal-level requirements are less, there is greater variation in signal level as the satellite moves from the horizon* position to near the zenith location. These signal-level changes can be up to 12 dB for LEOs and about 4 dB for satellites in MEO.

3.8 List of Symbols

F	Phasing parameter in Walker constellation.
h	Height of satellite orbit.
L_{FSL}	Free space loss, $20 \log (4\pi R_{slt}/\lambda)$.
n	Nadir angle.
N	Number of satellites in each plane.
r	Orbital radius $(R_e + h)$.
R_e	Earth's radius, 6378 km at equator.
R_{slt}	Slant range from satellite to ground transceiver.
S	Central angles to boundary of beam footprint.
T	Total number of satellites in the constellation.
T_s	Orbital period of satellite.
U	Users per cell.
α	Elevation angle of transceiver antenna beam.
$(\alpha + \beta) - \delta$	Spot beam width in elevation.
ψ	Nadir angle out to the horizon $(\alpha = 0)$.

---

*LEOs remain in view for approximately 15 min depending on the elevation angle, ground track, and altitude, and MEOs remain in view for about 2 h (for a 6-hour orbit).

Δ_1	Angular separation of corotating satellites.
Δ_2	Angular separation of counterrotating satellites.
ε	Angular one-half width of ground swath.
θ	Central angle.

References

Adams, W. S., and L. Rider, "Circular Polar Constellations Providing Continuous Single or Multiple Coverage Above a Specified Latitude," *The J. of the Astro. Sci.,* vol. 35, no. 2, April/June 1967.

Ballard, A. H., "Rosette Constellations of Earth Satellites," *IEEE Trans. AES,* September 1980, vol. 16, no. 5.

Benedicto, J., et al., "MAGSS-14: A Medium-Altitude Global Mobile Satellite System for Personal Communications at L-Band," *ESA Journal,* vol. 16, 1992.

Beste, D. C., "Design of Satellite Constellations for Optimal Continuous Coverage," *IEEE Trans. AES,* May 1978.

Burington, R. S., *Handbook of Mathematical Tables and Formulas,* Handbook Pub. Inc., Sandusky, Ohio, 1958.

Enara, H. E., et al., "Minimum Number of Satellites for Three-Dimensional Continuous Worldwide Coverage," *IEEE Trans. AES,* March 1977.

Hata, M., "Empirical Formula for Propagation Loss in Land Mobile Radio Service," *IEEE Trans. on Vehicular Technology,* August 1980.

Larson, W. J., and J. R. Wertz, *Space Mission Analysis and Design,* Kluwer Academic Pub. & Microcosm Inc., Dordrecht, Holland/Torrance, Calif., 1991.

Loral Cellular System Corp., "Authority to Construct a Low Earth Orbit Satellite System," (Globalstar), Application to the FCC, June 3, 1991.

Motorola Satellite Communications, Inc. (Iridium), "A LEO Mobile Satellite System," Application to the FCC, December 1980.

Motorola Satellite Communications, Inc., "Amended Application to Its Iridium System," File Nos. 9-DSS-P-91(87) CSS-91-010, December 1990.

Okamura, Y. E., et al., "Field Strength and Its Variability in VHF and UHF Land Mobile Radio Service," *Rev. Elec. Communications Lab.,* vol. 16, Sept.—Oct. 1968.

TRW, Inc., "Application for Radiodetermination/Mobile Satellite System," (Odyssey), Application to the FCC, June 3, 1991.

Walker, J. G., "Continuous Whole Earth Coverage for Circular Orbit Satellites," presented at the *IEEE Satellite Systems for Mobile Communications Conf.*, March 13–15, 1973.

4

Overview of the Frequency Bands Used in Non-GSO Satellite-Based Cellular Communications

4.1 Communications Channel Frequencies (Service Link)

The frequency management for mobile satellite service was addressed at both the WARC-87 and the WARC-92 (World Administrative Radio Conferences of 1987 and 1992, respectively). The frequency bands that were allocated to mobile satellite communications at the WARC-87 are indicated in Fig. 4.1. The bands were principally directed for use by satellites in geostationary orbit (GSO). No band allocations for satellites in nongeostationary orbits (non-GSO) were considered. The ITU defines non-GSO as satellites in either low earth orbit (LEOs), medium earth orbit (MEOs), or high elliptical orbit (HEOs). In the United States, American Mobile Satellite Corporation (AMSC) was initially licensed to operate a satellite network in GSO using the frequency bands 1545 to 1559 MHz (downlink) and 1646.5 to 1660.5 MHz (uplink). The service is to North America, Caribbean, and the coastal waters around North America. However, AMSC does not supply cellular service to the land masses, but the continent is covered by seven spot beams with a beam covering Mexico and the Caribbean. The spot beams permit a degree of frequency reuse.

AMSC provides generic mobile satellite service (MSS) to suitcase-size transceivers. The service is generic in that it provides land mobile-satellite service (LMSS), marine mobile-satellite service (MMSS), and aeronautical mobile-satellite service (AMSS). Partitioning the spectrum for these various services is indicated in Fig. 4.2. Subsequent to this licensing by the FCC, AMSC petitioned the FCC for additional spectrum contiguous to their original licensed bands, or the 28 MHz (14 + 14) already received. This was granted, and the new operating bands are 1530 to 1559 MHz (downlink) and 1626.5 to 1660.5 MHz (uplink). In addition, AMSC was allocated the bands 10.75 to 10.95 GHz and 13.0 to 13.15 GHz for their gateway feeder links.

At the 1992 WARC, additional frequency allocations were made for NGSO mobile-satellite service and radio determination satellite service (RDSS). Prior to 1992, no frequencies were allocated for use by NGSO systems. At the WARC-92, the United States sought allocations to enable voice-grade telephony for frequencies above 1 GHz and spectrum for frequencies below 1 GHz for satellites providing data and radio determination. In the United States, NGSO satellites providing voice were designated as big LEOs and those providing data and no voice as little LEOs, the former requiring a larger amount of spectrum.

The frequencies sought at WARC-92 are a primary MSS uplink allocation at 1610 to 1626.5 MHz, a secondary downlink allocation at 1613.8 to

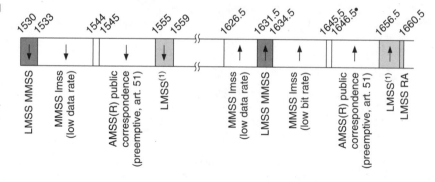

Figure 4.1

Frequency bands allocated to mobile satellite communications at the 1987 WARC.

(1): Footnote provision authorizes aircraft & ship communications via satellite.

• WARC-87 did not consider NGSO allocations.

1626.5, and a mating primary downlink allocation at 2483.5 to 2500 MHz. The United States was successful in this objective even though contentious due to other services using the L-band. This included the Russian use of part of the L-band spectrum for their GLONASS navigational system and also radio astronomy service at the lower part of the L-band.

The mobile frequency bands are indicated in Fig. 4.3. The frequencies at VHF and UHF are limited to NGSO MSS satellites and, in particular, in the United States to little LEOs. These bands are indicated in Fig. 4.4. As a condition for the allocation of approximately 3.5 MHz of spectrum, little LEO systems would share the spectrum with existing services. The most significant bands allocated were the RDSS band 1610 to 1626.5 MHz (L-band) and 2483.5 to 2500 MHz for NGSO services on a worldwide primary basis or to all ITU world regions 1, 2, and 3. In addition, part of the 1610 to 1626.5 MHz band was allocated on a secondary basis for bidirectional transmission. This is indicated in Figs. 4.3 and 4.5 as 1613.8 to 1626.5 MHz. In the United States, these bands are used to provide data and voice by NGSOs (big LEOs).

Because of the large number of LEO applicants vying for the big LEO spectrum, a problem developed on how to divide up the 16 GHz spectrum to accommodate all the applicants. In an attempt to resolve this problem, the FCC convened a negotiated rule-making (NRM) committee to seek a solution on frequency sharing. Several possible solutions were proposed by the various applicants, but none were acceptable to all parties. After several months of futile (but useful) negotiations, no agreement evolved and the meetings were terminated.

In order to break the impasse, the FCC proposed a solution by promulgating an NPRM (Notice of Proposed Rule Making) in June 1994 which segmented the band. This is shown in Fig. 4.6. The band is divided

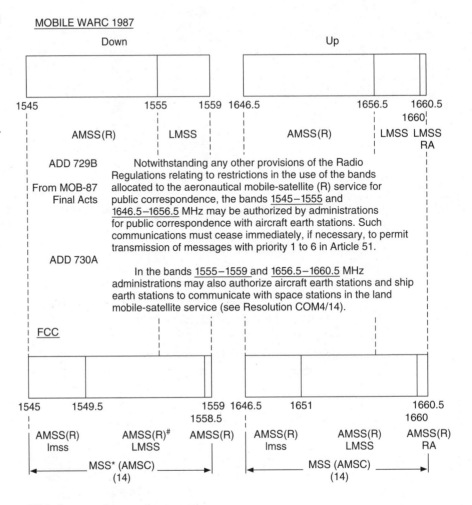

Priority use and preemptive access.
* Generic: AMSS, LMSS, MMSS

Note: Additional contiguous spectrum received by AMSC after these bands were licensed.

into two parts with the lower 11.35 MHz assigned to the systems using code division multiple access (CDMA). The upper 5.15 MHz was assigned to systems using frequency division multiple access (FDMA) and time division multiple access (TDMA). The only applicant in the latter category was Motorola's Iridium big LEO system. If more than one CDMA system is developed, the 8.25 MHz, which would have been assigned to a single user, would be expanded to 11.35 MHz. If the 3.10 MHz were to remain fallow (no additional CDMA system), then it can be used by Iridium as additional capacity is warranted. In the event that this plan was not

Figure 4.3

WARC-92 frequency allocation for mobile satellite service and radio determination satellite service.

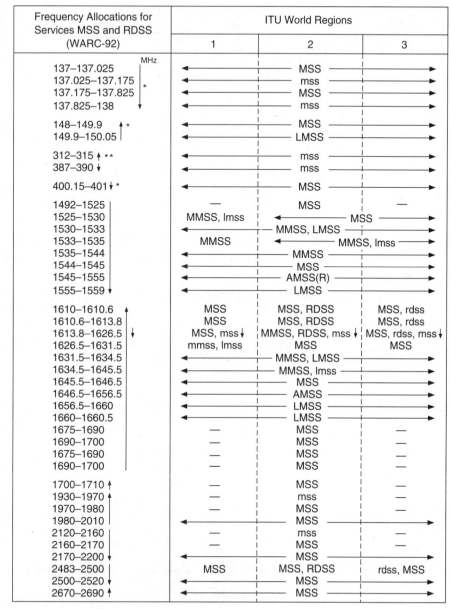

Frequency Allocations for Services MSS and RDSS (WARC-92)	ITU World Regions		
	1	2	3
137–137.025 MHz	◄——— MSS ———►		
137.025–137.175 *	◄——— mss ———►		
137.175–137.825	◄——— MSS ———►		
137.825–138	◄——— mss ———►		
148–149.9 ↑*	◄——— MSS ———►		
149.9–150.05	◄——— LMSS ———►		
312–315 ↑**	◄——— mss ———►		
387–390 ↓	◄——— mss ———►		
400.15–401 ↓*	◄——— MSS ———►		
1492–1525	—	MSS	—
1525–1530	MMSS, lmss	◄——— MSS ———►	
1530–1533	◄——— MMSS, LMSS ———►		
1533–1535	MMSS	◄——— MMSS, lmss ———►	
1535–1544	◄——— MMSS ———►		
1544–1545	◄——— MSS ———►		
1545–1555	◄——— AMSS(R) ———►		
1555–1559 ↓	◄——— LMSS ———►		
1610–1610.6 ↑	MSS	MSS, RDSS	MSS, rdss
1610.6–1613.8	MSS	MSS, RDSS	MSS, rdss
1613.8–1626.5 ↓	MSS, mss↓	MMSS, RDSS, mss↓	MSS, rdss, mss↓
1626.5–1631.5	mmss, lmss	MSS	MSS
1631.5–1634.5	◄——— MMSS, LMSS ———►		
1634.5–1645.5	◄——— MMSS, lmss ———►		
1645.5–1646.5	◄——— MSS ———►		
1646.5–1656.5	◄——— AMSS ———►		
1656.5–1660	◄——— LMSS ———►		
1660–1660.5	◄——— LMSS ———►		
1675–1690	—	MSS	—
1690–1700	—	MSS	—
1675–1690	—	MSS	—
1690–1700	—	MSS	—
1700–1710 ↑	—	MSS	—
1930–1970 ↑	—	mss	—
1970–1980	—	MSS	—
1980–2010	◄——— MSS ———►		
2120–2160	—	mss	—
2160–2170	—	MSS	—
2170–2200 ↓	◄——— MSS ———►		
2483–2500	MSS	MSS, RDSS	rdss, MSS
2500–2520 ↓	◄——— MSS ———►		
2670–2690 ↑	◄——— MSS ———►		

* The use of these bands is limited to non-GSO satellite systems.
** May also be used for non-GSOs.
• The bands indicated are for communication services.
• Many of the bands are also allocated to other services.
• Feeder links are not shown for microwave operation, but for bands greater than 1 GHz; the feeder frequencies are located in the higher microwave bands, typically C-band, Ku-band, or Ka-band.
• Further conditions and information can be found in Final Acts WARC-92.

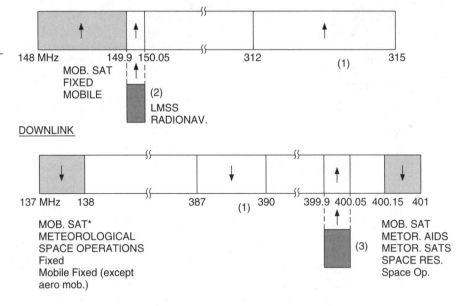

Figure 4.4
Frequencies available
for little LEOs
(<1 GHz).

: Bands planned to be used by U.S. applicants.

Notes: (1) May be used both by LEO and GSO systems for mss on a noninterference to
           other services (FN 641, 641A), all ITU regions. At the present time, no plans to be
           used by U.S. small LEO applicants.
       (2) May be used on a secondary basis until Dec. 1996. Now being used by TRANSIT
           (will be phased out (supplanted by GPS)). Coordination required.
       (3) Allocated to MSS on a primary basis after 1.1.1997, nonvoice, non-GSO.
           *Secondary in bands: 137.333–137.367, 137.485–137.515, 137.605–137.635, and
           137.753–137.787 MHz until Jan. 1, 2000.

acceptable to the applicants, an auction or lottery would be considered to
parcel out the spectrum: a route not favored by the applicants.

4.2 Feeder Link Channel Frequencies

4.2.1 Ka-Band Spectrum Allocation

Several frequency bands are normally used for operation of a satellite system. One is the communication frequencies (service link) which are used
to communicate between the user and the satellite. For little LEOs, this is

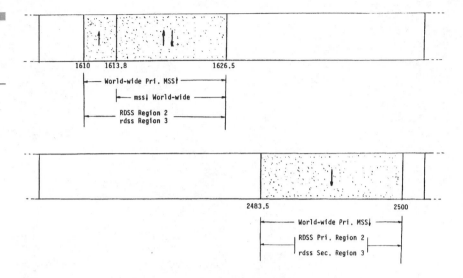

Figure 4.5
WARC-92 allocated big LEO MSS service bands (>1 GHz).

in the VHF/UHF band (see Fig. 4.4). For big LEOs, this band is at L-band (1.6 GHz) and S-band (2.5 GHz). The second frequency band is for telemetry tracking and command (TT&C) and is thereby used to determine the status of the satellite, commanding the various functions on the satellite, and returning status information to earth. This function may be performed by using frequencies in the service band if spectrum is available. If not, another band must be used. The third band required is the feeder link band, which is the link between the satellite and the local telephone network (PSTN). The interface between the PSTN and the satellite is the gateway earth station. There are several bands which can be used for this link. For small LEOs, as indicated previously, this will be performed in the communication bands. For big LEOs or GSO mobile-satellite service, the bands being considered are C-band (5 to 6 GHz), Ku-band (15 GHz), and Ka-band (18 to 30 GHz).

There is another link used by the Iridium system. That is, the intraorbit and/or interorbit satellite link. This permits communications between satellites in the same or in adjacent orbits. This mode of operation reduces the number of gateways required globally and connects to the closest PSTN nearest the intended destination. The Iridium system employs the frequency range 23.18 to 23.38 GHz for the space-to-space communications cross-links. A scenario depicting the operation of a LEO/MEO system and links employing the bands alluded to earlier is shown in Fig. 4.7.

WARC-92 allocated service link spectrum, but not feeder link spectrum. This issue was addressed at the WRC-95 (World Radio Conference, 1995). On the agenda at the WRC-95 (held in November 1995) was alloca-

Figure 4.6
Big LEO FCC notice
of proposed rule
making.

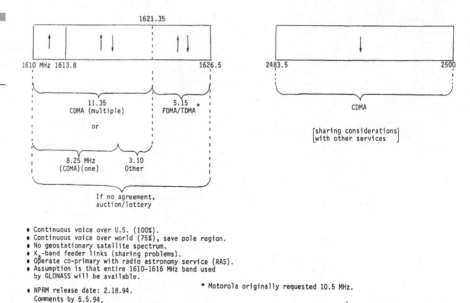

Figure 4.6
Big LEO FCC notice of proposed rule making.

tion for the feeder links for non-GSO MSS satellites. In addition, the United States sought spectrum at Ka-band to be used by the non-GSO fixed satellite service (FSS). Although not originally an issue or on the agenda at the WRC-95, nevertheless, the United States was successful in getting allocation in this band for FSS. American big LEOs are designed to operate in three different bands (C, Ku, and Ka), and the United States sought spectrum in these bands.

The United States went to the conference with a proposal seeking to obtain a 500-MHz (in both directions) spectral swath for non-GSO FSS and 400-MHz (in both directions) bandwidth for non-GSO MSS at Ka-band. The Ka-band proposal prior to the WRC-95 is indicated in Fig. 4.8. The 500 MHz for non-GSO for FSS was a contentious issue, but the United States was able to obtain 400 MHz allocated for this service. The significant issue was feeder frequencies for the non-GSO MSS. The aim was 400-MHz bandwidth of spectrum, but 300 MHz was allocated. The consensus among the concerned LEO participants is that this amount of spectrum is adequate to serve the needs of Motorola's Iridium and TRW's Odyssey satellites which sought the spectrum. They can now complete their systems as originally designed. The other 100 MHz (100 MHz up, 100 MHz down) for both FSS and MSS is proposed for the agenda for the WRC-97. The Ka-band spectrum allocated at the WRC-95 is shown in Fig. 4.9. The speckled and hatched flanking areas indicate the feeder uplink frequencies for the non-GSO MSS satellites (Iridium, Odyssey).

Figure 4.7
Non-GSO satellite
system operational
scenario.

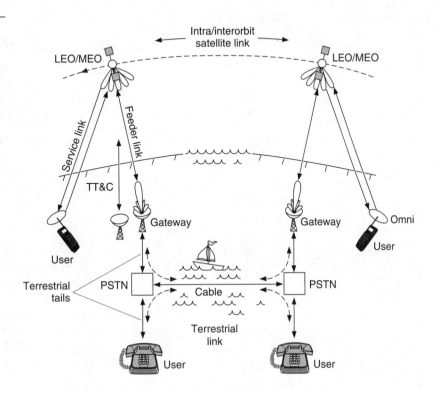

Figure 4.7
Non-GSO satellite system operational scenario.

- For small LEOs, the satellite beam is omnidirectional.
- For big LEOs, the satellite will use spot beams.
- The service link is at VHF/UHF (small LEOs), (L/S)-band (big LEOs).
- The feeders are at Ka-band, Ku-band, and C-band.
- The cross-link frequencies are in Ka-band (Iridium).

The results obtained at the WRC-95 for the Ka-band were satisfactory for the United States. However, indigenously, the spectrum was still contentious principally between the terrestrial groups with their LMDS (Local Multipoint Distribution System) and the satellite proponents. Several meetings were held concerning the partitioning of the spectrum among the various groups. Several plans evolved, but none were completely satisfactory to all parties. At this juncture, the FCC considered all input and arrived at a cogent segmentation plan which was satisfactory to all participants, and their concerns were primarily satisfied as well.

The FCC's segmentation plan resulted in the First Report and Order which was released on July 22, 1996. Appended to the R&O were rule

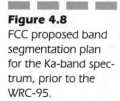

Figure 4.8
FCC proposed band segmentation plan for the Ka-band spectrum, prior to the WRC-95.

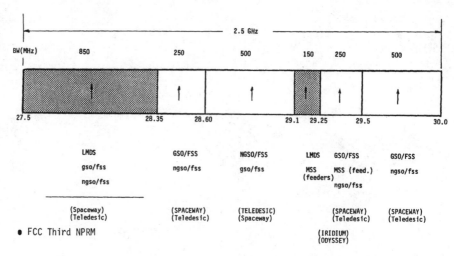

N.B.: Note primary service (caps) and sceondary service (lower case).

Spaceway: Hughes proposed global FSS satellite communication system operating at GSO. Vying for K_a-band spectrum.
Telesesic: A proposed global FSS satellite communication system at LEO. Vying for K_a-band spectrum.
LMDS: Local Multi-Point Terrestrial Distribution System providing TV to the home. Vying for K_a-band spectrum.

amendments to volume 47 of the Code of Federal Regulations (CFR) and Parts 25 and 101 of the commission rules. In addition, this was supplemented by a Fourth Notice of Rulemaking (4th NPRM) which proposed to allocate additional spectrum in the band 31.0 to 31.3 GHz for use by Local Multipoint Distribution Systems (LMDS).

The segmentation plan indicated in the R&O is shown in Fig. 4.10. Note that the MSS feeder link bands are comparable to those allocated at the WRC-95 (see Fig. 4.9), save a 100-MHz sliver (29.4 to 29.5 GHz) and 100 MHz (19.6 to 19.7 GHz) on the downlink, which were not allocated, but the United States will seek these segments at the WRC-97. The bands will be used by Motorola for their Iridium LEO system and TRW for their Odyssey MEO system. However, neither will need the full 400 MHz and sharing does not appear to be a problem.

The bands 29.1 to 29.5 GHz and 19.3 to 19.7 GHz are paired bands for the non-GSO feeder links. These are shared with other services, but any problem of interference will be tractable.

The LMDS proponents are getting the 1000 MHz of spectrum they were seeking even though it is not contiguous and part of it is shared by the MSS feeder link on a primary basis. There is additional spectrum which they may receive in the band 31.0 to 31.3 GHz (see Fig. 4.10) for technical flexibility. This decision and contending issues are pending and are

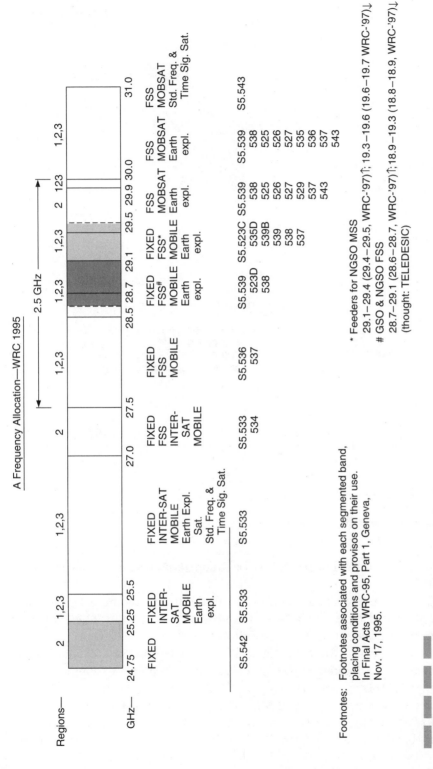

Figure 4.9 WRC-95 Ka-band frequency allocation.

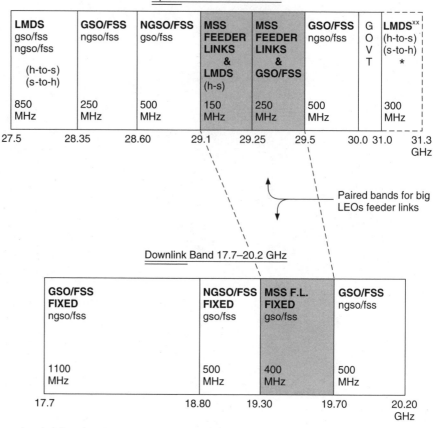

Uplink Band 27.5–30.0 GHz

Downlink Band 17.7–20.2 GHz

- h-s: hub to sub.
- s-h: sub. to hub.

Source: FCC 4th R&O, CC Docket #92-297 (7.22.96)
        * 4th NPRM (7.22.96)

the subject of the Fourth Notice of Rulemaking which was released with this R&O.

Three bands were allocated for the feeder linksat by the WRC-95. These are indicated in Fig. 4.11: C-band, Ku-band, and Ka-band. The applicants affected are shown in the third column.

In addition to the frequencies shown in Fig. 4.4 for the small LEOs, which were allocated at the WRC-92, the United States sought an additional 6 MHz of spectrum below 1 GHz on a global basis. *The United States was only able to obtain 2 MHz of spectrum (455 to 456 MHz and 459 to 460 MHz) for use in region 2 only.* The concern of the conference attendees was that the lower bands are used for terrestrial services worldwide. Inter-

Figure 4.11
C-band, Ku-band, and
Ka-band MSS feeder
link frequencies.

Bands	Transmission Direction	United States Applicant
5091—5250 MHz↑	uplink	Globalstar
6700—7075↓	downlink	ICO Global Communications
15.40—1545 GHz↓	downlink	Mobile Communications
15.45—15.65↕	bidirectional	Holding Inc. (MCHI)
15.65—15.70↓	downlink	(ELLIPSO)
19.3—19.6 GHz↑↓*	downlink/uplink	Iridium
29.1—29.4↑*	uplink	Odyssey

Source: Final Acts WRC-95.

• In some of these bands, restrictions apply.

*Additional spectrum (19.6—19.7 GHz) and (29.4—29.5 GHz) will be on the agenda for the WRC-97 conference.

ference to terrestrial services is the issue here. Administrations not supporting the U.S. proposals did so as a result of what they felt was a lack of information on how the proposed bands be shared between LEO systems and existing services in their own countries. However, the United States will put this on the WRC-97 agenda for additional spectrum allocation. The motive for this is that the United States has several new applicants (second round) seeking spectrum in the VHF/UHF range in addition to that which is already utilized by the first round little LEOs: ORBCOMM, VITA, and GE Starsys. There may not be enough WRC-92 allocated spectrum to accommodate more than one new entry.

Two main LEO issues to be taken up at the 1997 conference will be: (1) the additional 100 MHz of spectrum for both the MSS feeders and the non-GSO fixed satellite service, and (2) additional spectrum below 1 GHz for the little LEOs.

In 1994, the FCC received eight additional applicants (second round) seeking spectrum for little LEO operation. The eight new applicants include: CTA Commercial Systems, Inc. (CTA); E-Sat; Final Analysis; GE American Communications; LEO One USA; ORBCOMM; Starsys; and VITA. The last three in this grouping have also received spectrum in the first round.

In October 1996, the FCC promulgated an NPRM proposing that the existing spectrum allocated for little LEOs may be able to accommodate up to three additional systems. One technique being considered to share this spectrum is time sharing, realizing that satellites will be in view for short periods of time during their orbital periods. However, three from the first round proceedings (ORBCOMM, Starsys, and VITA) were excluded from receiving any additional spectrum because of their success in receiving spectrum in the first round proceedings. GE Americon may also be in this category, since it has an interest in the first round Starsys.

4.2.2 S-Band (2-GHz) MSS Spectrum Allocations

At the WRC-95 additional MSS spectrum was allocated in the S-band and proposed to be used by a relatively new entry in the big LEO arena, that is, ICO Global Communications which plans to put 10 satellites into medium earth orbit. ICOGC is an offshoot of Inmarsat. The constellation will be operating in two inclined circular planes and using service links in the 2-GHz band. The S-band allocations are indicated in Fig. 4.12. The exact frequencies which will be used have not been specified (as of 6/96), but the 2-GHz spectrum is expected to be available by the year 2000: a decision made at the WRC-95.

A summary of the frequency bands to be used by the big LEOs in both their service bands and feeder bands is indicated in Fig. 4.13.

4.3 Arrows on Bands

$\updownarrow$ Bidirectional.

$\leftrightarrow$ Cross-link.

$\downarrow$ Downlink.

$\uparrow$ Uplink.

Figure 4.12
WRC-95 S-band
allocations.

1990—2010 MHz$\uparrow$	Globally MSS$\downarrow$	available Jan. 2000
2170—2200$\downarrow$	Globally MSS	
1980—1990$\uparrow$	Regions 1 & 3, Region 2 available year 2000	save U.S., Canada due to PCS
2010—2025$\uparrow$	Region 2 MSS	available starting in 2005
2160—2170$\downarrow$	Region 2 MSS	U.S./Canada year 2000

Figure 4.13
Frequencies to be used by big LEOs providing satellite-based cellular systems.

System	Communication Service Bands	Feeder Bands
Globalstar	2483—2500 MHz↓	5091—5250 MHz↑↓
	1610—1618.25↑ (BW: 8.25 MHz)[1]	6700—7075↓
	1610—1621.5↑ (BW: 11.35 MHz)[2]	
Iridium	23180—23380 MHz	29.1—29.4 GHz↑*
	1621.35—1626.5↑↓ 1618.265—1621.35↓↑ (BW: 3.1)[3] cross-link**	19.3—19.6↓*
Odyssey	2583—2500↓	29.1—25.4 GHz↑*
	1610—1618.25↑ (BW: 8.25)[1] 1610—1621.5↑ (BW: 11.35)[2]	19.3—19.6↓*
ICO	S-Band 2 GHz	5/7 GHz

Source: For U.S. satellites, NPRM, Jan. 19, 1994.
*29.4—29.5 GHz: On agenda for WRC-97.
*19.6—19.7 GHz: On agenda for WRC-97.
**Iridium will also have a cross-link frequency band for intra- and interorbit communications, 23.18—23.38 GHz.
(1) For one CDMA system.
(2) For multiple CDMA systems.
(3) Possible additional spectrum for TDMA systems if not used by CDMA systems.

References

FCC Big LEO Notice of Proposed Rule Making, January 1994.

FCC First Report and Order, CC Docket #92-297, and Fourth Notice of Proposed Rule Making, July 22, 1996.

FCC NPRM, IB Docket No. 96-220, Amendment to Part 25 of the Commission Rules to Establish Rules & Policies Pertaining to the Second Processing Round of the Nonvoice, Nongeostationary Mobile-Satellite Service, October 29, 1996.

Final Acts of the World Administrative Radio Conference (WARC-92), Malaga-Torremolinos, Spain, 1992.

Final Acts of the World Administrative Radio Conference for the Mobile Services (MOB-87), Geneva, 1987.

Final Acts, World Radio Conference (WRC-95), Part 1, Geneva, 1995.

Report of the MSS Above 1 GHz Negotiated Rule Making Committee, April 6, 1993.

Propagation Anomalies and Signal Statistical Distributions in Wireless Communications

5.1 Propagation Loss in Mobile Wireless Systems

It is generally known that the classical Friis equation for propagation loss in a mobile cellular environment is inaccurate. That is, propagation losses occur which are higher than the K/R^2 relationship. Hata has developed an empirical formula for propagation loss for systems operating in the 1-GHz range. In Hata, this is given as:

$$L_p(\text{dB}) = 69.55 + 26.16 \log f_c - 13.82 \log h_b - a(h_m)$$
$$+ (44.9 - 6.55 \log h_b) \log R \qquad \text{[5.1]}$$

where h_b = base station height, 30 to 200 m
f_c = 150 to 1500 MHz
R = range, 1 to 20 km
$a(h_m)$ = correction factor for vehicle antenna height

We notice that the fifth term (so-called factor B by Hata) in the right member is the only one that contains the range factor R. It is clear that the exponential on R is a function of the base station antenna height. The higher the base station antenna, the smaller the exponential becomes. For example, for a height of 70 m, the exponential of $R \approx 3.3$ and for a 30-m height, $R^{3.5}$. For a height between 5 to 7 m, the value is R^4. The loss is not 6 dB per octave change in range but more like 12 dB per octave change in range. Experimental results, and those submitted to FCC in applications for licenses, confirm this loss trajectory as a function of range. Figure 5.1 from Hata's paper supports these results.

In mobile wireless radio systems the signal received, either from a terrestrial system or a satellite-based network, will be subjected to propagation anomalies. This results from various terrain features. In addition, other perturbations can be due to atmospheric and ionospheric effects.

Under ideal conditions there would be line-of-sight propagation, and a coherent signal would be received. In this situation, the basic Friis equation for received signals would prevail. That is, the signal is attenuated by the $1/R^2$ relationship where R is the range between the transmitter (satellite or base station) and an in-the-clear receiver. The signal is therefore increased or decreased by 6 dB by an octave change in range.

However, in real life, this simplistic scenario does not exist where the environment is hostile, and the received signal will generally be reduced from the ideal case. The signal received is no longer deterministic but

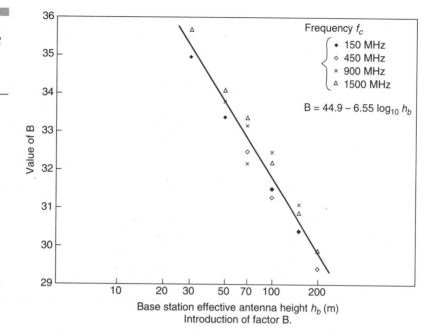

Figure 5.1

The exponential of R (factor B) in Hata's propagation loss equation.

Introduction of factor B.

now assumes a statistical character. A fixed-signal level has no meaning, and the levels received must be dealt with in probabilistic terms. Hostile effects which exist in a signal channel due to external factors include the following:

■ Ionosphere causing Faraday rotation of the signal RF vectors

■ Atmospheric conditions resulting in signal attenuation and possible polarization rotation. (The higher frequencies are plagued with rain-induced signal attenuation.)

■ Shadowing and/or blockage reducing direct-signal levels or, in the extreme, complete blocking of the signal

■ Multipath and specularly reflected signals

■ Doppler shift requiring adaptive or frequency tracking techniques

■ Internal factors such as nonlinearities, interchannel interference (ICI), and adjacent channel interference (ACI) (not included here)

The relative importance of the effects listed previously depends on several factors such as frequency of operation, mobile location and velocities, diurnal and seasonal variations of the atmosphere, and, of course, system parameters such as antenna properties or receiver characteristics which may help to discriminate against these anomalies.

In the satellite-based mobile system, the direct downlink wave from the satellite passes through the ionosphere. Here there is a rotation of the electric vectors as they propagate through this medium. A linearly polarized electric vector can be broken down into two orthogonal components. Each vector is subjected to a different phase velocity, and, upon exiting the layer, the composite wave will be rotated from the direction at which it entered. This rotation is given by the following expression.

At high frequencies (nominally above 100 MHz), Faraday rotation along a path is given by Flock:

$$\phi = 2.36 \times 10^4/f^2 \int NB \cos \theta_B \, dl \qquad \text{radians} \qquad \text{[5.2]}$$

where ϕ = Faraday rotation in radians

f = frequency in Hertz

l = length through the ionosphere

N = electron density in electrons/m^3, the integral of the electron density is weighted by the value of $B \cos \theta_B$ along a path. That is, the number of electrons along a path in a column 1 m^2 in cross section.

θ_B = angle between the direction of propagation and the earth's magnetic-flux density vector (B varies with the cube of the radius from the earth's center and has a low value above $\approx$2000 kilometers.)

B = magnetic flux density of the earth's field.

For LEOs, it is sufficiently accurate to replace $B \cos \theta_B$ by an average value B_L and it can be placed outside the integral.

$$\phi = 2.36 \times 10^4/f^2 \int N \, dl \qquad \text{[5.2']}$$

The total rotation can be seen to vary inversely with f^2 and to be equal to the integral of electron density weighted by the value of $B \cos \theta_B$ along the path. The integral stands for total ionospheric electron content (TEC) along the path in electrons per meter square (el/m^2). One therefore obtains

$$\phi = (2.36 \times 10^4/f^2)\text{TEC} \qquad \text{[5.2'']}$$

The significance of this is where Faraday rotation occurs (100 MHz), a linearly polarized wave will be subjected to rotation of its vector creating a polarization loss at the receiver.

Representative values of Faraday rotation as a function of frequency and parametric TEC (el/m^2) is shown in Fig. 5.2. TEC is not fixed and varies seasonally and diurnally.

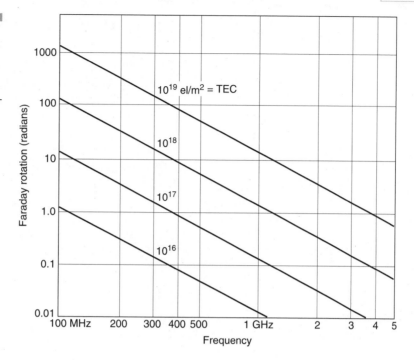

Figure 5.2
Faraday rotation of a linearly polarized wave negotiating the ionosphere.

The polarization loss may be given by

$$L_p = 20 \log \phi \qquad \text{dB} \qquad\qquad [5.3]$$

For a LEO operating frequency of approximately 1.2 GHz, and using the extreme value of TEC = 1.86×10^{18} el/m^2 and $B = 0.43 \times 10^{-4}$ Webers/m^2 (CCIR, 1986b, Report 263-6), the polarization rotation is about 63°. Observe from Fig. 5.2 that at the lower frequencies the electric vector can undergo several 360° rotations.

Rainfall is not considered a serious problem for mobile systems operating at VHF or at the lower microwave frequencies and up to about 10 GHz. At the high microwave frequencies, the droplet size (oblate in shape) becomes significant in comparison to the wavelength and produces greater attenuation. Also, some depolarization can occur for signals which contain both vertical and horizontal electric vectors. Raindrops are not spherical but shaped like oblate spheroids, and the horizontal wave is attenuated more than the vertical vector. This can cause depolarization effects in addition to attenuation.

The tropospheric effects which influence the signal from a satellite are predominantly due to rain, clouds, and gases. In all cases, the attenuation increases with frequency, and the rain losses predominate.

5.2 Mobile Signal Reception

In terrestrial cellular and/or satellite-based mobile systems, the signals reaching the receiver originate from several different directions even though the transmitter is a single source. A diagram depicting the origination of signals reaching the receiver is shown in Fig. 5.3. Notice that the signal reaching the receiver is the sum of several signals coming from different directions. In the simplest situation, a signal such as r_1 is a direct wave with no intervening obstacles, save possibly the disturbances caused by the ionosphere and/or the atmosphere. This is the line-of-sight wave or direct wave. The transmission basically follows

Figure 5.3
Propagation model in wireless communications.

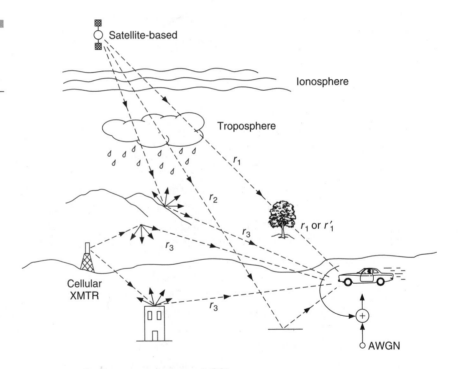

- r_1 = direct component of wave (LOS)
- r_2 = specularly reflected wave (generally $r_2 < r_1$)
- r_1, r_2 = together generally referred to as coherent components
- r_3 = diffused random component, Rayleigh statistics
- $r_1 + r_3$ = mixture Rayleigh + dominant = Nakagami/Rician statistics
- r_1' = if r_1 is shadowed, lognormal statistics
- AWGN = of course, additive white gaussian noise is ubiquitous

the Friis relationship where the signal is attenuated by the $1/R^2$ factor, where R is the range between the transmitter and the receiver. The direct wave is deterministic (implying no random components) and coherent.

The r_2 wave results in a specular reflection which eventually reaches the receiver along with the direct wave which can be in phase or anti-phase with the direct wave. As the name suggests, the wave(s) are reflected from a relatively smooth surface. The reflection coefficient depends on several factors: the frequency of operation and the conductivity and relative permittivity of the terrain. This information is available in ITU CCIR Report 527-1. The reflected signal is actually a delayed and shifted replica of the direct wave and generally considerably smaller than the direct wave. Nonetheless, it is desirable to suppress it as much as possible since it can subtract or add to the direct component. Another mitigating factor is that the reflected wave illuminates the receiver antenna beam in its low-gain region (low-elevation angles). In addition, the rays do not go through the phase center of the antenna beam. In studies, contributions to the signal received are therefore usually neglected.

The other component which affects the purity of the signal received is the fading diffused component r_3. The length of this vector results from the summation of a multiplicity of comparable* random vectors which are scattered off the earth. The summation of random vectors is depicted in Fig. 5.4. The component assumes a statistical fading envelope. The envelope assumes a Rayleigh distribution and will be discussed in greater detail later. The number of scatterers required for this distribution is about six. Other distributions can occur depending on the terrestrial environment. A classification of the more typical distributions is listed in Table 5.1. These, too, will be indicated in subsequent sections along with their properties.

Most distributions degrade the performance of a system, except the nonfading classification. Varying degrees of fading margin (dB) will be necessary to realize the desired performance or service for X percent of the time. In digital systems one can expect degradation in bit error rate (BER). However, margins can be more easily controlled in terrestrial systems than in satellite-based systems since power (or EIRP) is at a premium in spacecraft.

---

*Scatterers which are comparable in magnitude in real life may be unrealistic, but we will not argue this point here.

Figure 5.4
Summation of random signal vectors and constant vector in a mobile wireless environment.

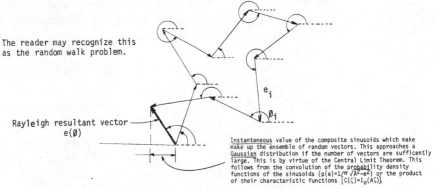

The reader may recognize this as the random walk problem.

Rayleigh resultant vector e(Ø)

e_i

$Ø_i$

Instantaneous value of the composite sinusoids which make up the ensemble of random vectors. This approaches a Gaussian distribution if the number of vectors are sufficiently large. This is by virtue of the Central Limit Theorem. This follows from the convolution of the probability density functions of the sinusoids $(p(e)=1/\pi\sqrt{A^2-e^2})$ or the product of their characteristic functions $[C(\zeta)=I_0(A\zeta)]$.

RAYLEIGH DISTRIBUTION: The sum of a large number of random vectors e_i with random relative phase (Ø)* is a vector e(Ø) which is Rayleigh distributed in amplitude, and phase is uniformly distributed. Number of phasors have to be sufficient (>≈6-7).
      * All equally likely.

(a)

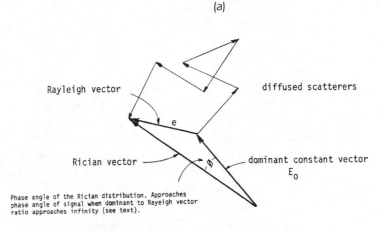

Rayleigh vector

diffused scatterers

e

Rician vector

dominant constant vector E_0

Ø

Phase angle of the Rician distribution. Approaches phase angle of signal when dominant to Rayeigh vector ratio approaches infinity (see text).

RICIAN DISTRIBUTION: Distribution of the length of a vector which is the sum of a fixed vector (E_0) and a vector "e" whose length has a Rayleigh distribution. If the Rayleigh vector has a most probable value of "0", and "A" is the length of the fixed vector we obtain the Nakagami/Rician distribution

(b)

5.3 The Rayleigh Distribution

The Rayleigh distribution has found wide application in many areas of science and technology. The distribution first came to light in Lord Rayleigh's publication entitled *The Theory of Sound*, vol. 1, published about 1880. Also, application is found in radar where backscatter from aircraft (the skin is an ensemble of random scatterers) assumes a Rayleigh distribution. Of course, this distribution is also found in cellular com-

TABLE 5.1

A Propagation
Medium
Classification

- *Nonfading.* Steady unperturbed signal follows the Friis equation where propagation loss follows the k/R^2 relationship and where k is an absorptive* constant, and R is range.

- *Rayleigh fading.* Combining random scatterers in amplitude and phase. No direct component, signal strength follows the Hata model or less (k/R^4) in terrestrial cellular systems.

- *Nakagami-Rician.* Distribution resulting from the composite of diffused scatter and steady signal.

- *Lognormal distribution.* Direct signal received where transmission has been shadowed by foliage, etc.

- *Suzuki Distribution.* Mixed shadowed and diffused signal.

- *Nakagami-m distribution.* Generalized distribution can evolve into any of the other distributions depending on parameters.

- *Complete blockage.* Signal dropout (complete outage).

---

*Contains the system parameters.

munications. Since this distribution is found in many areas of technology, it may be of interest to discuss some of its attributes before we continue.

The trajectory of the Rayleigh distribution for parametric σ is shown in Fig. 5.5.

Figure 5.5
March of the Rayleigh probability density functions and probability integrals.

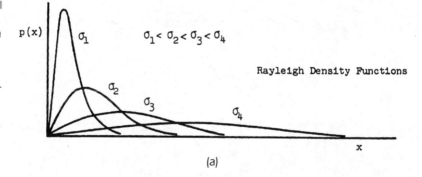

(a)

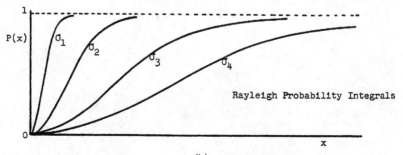

(b)

A graph of the probability density functions is shown in Fig. 5.5*a*. The cumulative distribution functions are shown in Fig. 5.5*b*. The latter is the integral of the density function as shown in Eq. [5.5].

The amplitude of the diffused signal envelope at the receiver undergoes fading and is represented by the well-known probability density function expression and uniquely specified by the variance σ^2.

$$p(v) = (v/\sigma^2) \exp(-v^2/2\sigma^2) \qquad 0 \le v < \infty \qquad [5.4]$$

The cumulative distribution function is

$$P(v_0) = \text{prob}\,[v \le v_0] = \int_0^{v_0} p(v)\, dv = 1 - \exp(-v_0/2\sigma^2) \qquad 0 \le v < \infty \qquad [5.5]$$

A tabulation of this cumulative distribution is shown in Table 5.2.

If we start at the other end, the probability that the envelope (say) exceeds some specified value v_0 is obtained by integrating probability density function Eq. [5.6] from v_0 to infinity.

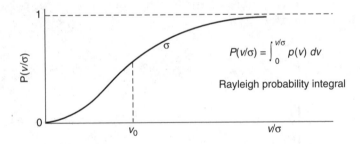

$$P(v/\sigma) = \int_0^{v/\sigma} p(v)\, dv$$

Rayleigh probability integral

$$\text{prob}\,[\text{envelope} > v_0] = \int_{v0}^{\infty} (v/\sigma^2) \exp(-v^2/2\sigma^2)\, dv$$

$$= \exp(-v_0^2/2\sigma^2) \qquad [5.6]$$

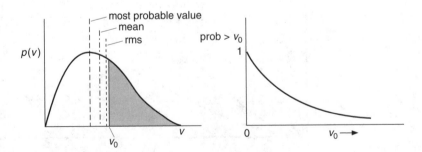

TABLE 5.2

Tabulation of the Rayleigh Cumulative Distribution

v_0	$P(v)$	v_0	$P(v)$
0.0	0.0000	2.5	0.9561
0.1	0.0049	2.6	0.9650
0.2	0.0198	2.7	0.9739
0.3	0.0440	2.8	0.9801
0.4	0.0769	2.9	0.9851
0.5	0.1175	3.0	0.9889
0.6	0.1647	3.1	0.9918
0.7	0.2173	3.2	0.9940
0.8	0.2738	3.3	0.9957
0.9	0.3330	3.4	0.9969
1.0	0.3935	3.5	0.9978
1.1	0.4539	3.6	0.9985
1.2	0.5133	3.7	0.9989
1.3	0.5704	3.8	0.9992
1.4	0.6247	3.9	0.9995
1.5	0.6754	4.0	0.9997
1.6	0.7220	4.1	0.9998
1.7	0.7642	4.2	0.99995
1.8	0.8021	4.3	0.999957
1.9	0.8355	4.4	0.999959
2.0	0.8645	4.5	0.99996
2.1	0.8898		
2.2	0.9110		
2.3	0.9290		
2.4	0.9439		

The moments about the origin are designated by

$$m_v = \int_0^\infty v^n p(v)\, dv \qquad [5.7]$$

The first moment (mean = average value or expectation) is given as

$$E(v) = \bar{v} = \int_0^\infty vp(v) = \int_0^\infty v(v/\sigma^2)\exp\left(-v^2/2\sigma^2\right) dv$$

$$= \sqrt{\pi/2}\,\sigma \qquad [5.8]$$

The second moment (mean square value) is

$$E(v^2) = \overline{v^2} = \int_0^\infty v^2 p(v)\, dv = 2\sigma^2 \qquad \text{mean square value} \qquad [5.9]$$

The root mean square value is equal to $\sqrt{2}\sigma$. The moments are about zero, but more frequently the moments are taken about the mean. Therefore, the first centroidal moment is

$$u_1 = \int_0^\infty (v - \bar{v})p(v)\, dv = 0 \qquad [5.10]$$

and the second centroidal moment (*variance*) is

$$[E(x - \bar{x})^2] = u_2 = \int_0^\infty (v - \bar{v})^2 p(v)$$

$$= \int v^2 p(v)\, dv - 2\bar{v}\int vp(v)\, dv + \bar{v}^2 \int p(v)\, dv \qquad [5.11]$$

but the first term is the *second* moment about the origin equal to $\overline{v^2}$, the second integral is the *first* moment about the origins equal to $\bar{v}$, and the third *integral* is equal to 1.

We therefore have the variance

$$\tau = u_2 = \overline{v^2} - \overline{2v^2} + \bar{v}^2 = \overline{v^2} - \bar{v}^2$$

$$= 2\sigma^2 - (\pi/2)\sigma^2 = \text{mean square} - \text{square of the mean} \qquad [5.12]$$

The standard deviation is therefore

$$\tau = \sqrt{2 - (\pi/2)}\,\sigma \qquad [5.13]$$

In terms of power, we have ac power + dc power (into one ohm resistor), that is, adding the variance to the mean square:

$$v^2 = u_2 + v^2 = (2 - (\pi/2))\sigma^2 + (\pi/2)\sigma^2 \qquad [5.14]$$

$$= 2\sigma^2 \qquad \text{(total power or power in diffused component, also mean signal power)}$$

Occasionally the median value is used and is defined where $P(v_m) = 0.5$. From Eq. [5.5] we obtain

$$1 - \exp(-v_m^2/2\sigma^2) = 0.5 \qquad [5.15]$$

Therefore

$$\exp(-v_m^2/2\sigma) = \tfrac{1}{2} \qquad [5.16]$$

or

$$v_m = (2 \ln 2)^{1/2}\sigma$$

Gathering the various parameters we have the results shown in Table 5.3.

The phase of the Rayleigh random vector is a random variable and has a uniform distribution. Intuitively, this makes sense since the phases have no inclination to assume any localized density.

However, to put phase and the Rayleigh amplitude distribution on a more mathematical footing, the Rayleigh density function can be derived by assuming that the resultant on n vectors can be decomposed into two independent variates (random variables) x and y. These are gaussian distributed with equal standard deviation and zero mean $(0, \sigma)$. Since the

TABLE 5.3

Parameters of the Rayleigh Distribution

Parameter	Value
Mode	σ
Median	$(2 \ln 2)^{1/2}\sigma = 1.18\sigma$
Mean	$\sqrt{\pi/2}\,\sigma = 1.25\sigma$
Root mean square (rms)	$\sqrt{2}\,\sigma = 1.41$
Variance	$(2 - (\pi/2))\sigma^2 = 0.43\sigma$
Standard deviation	$\sqrt{(2 - (\pi/2))}\,\sigma = 0.655\sigma$

variables are independent, their joint probability density function is therefore

$$p(x, y) = p(x)p(y)$$
$$= (1/\sigma \sqrt{2\pi})^2 \exp[-(x^2 + y^2)/2\sigma^2] \qquad \textbf{[5.16a]}$$

The variables are changed in order to obtain distribution of amplitude and phase.

The new joint distribution is now given as $q(R, \theta)$. To go from the original variates x, y to the new variates R, θ requires the exercise of the Jacobian determinant (see Rice and Miller in Reference section).

$$J\left(\frac{x, y}{R, \theta}\right) = \left\|\frac{\partial(x, y)}{\partial(R, \sigma)}\right\| = \begin{Vmatrix} \dfrac{\partial x}{\partial R} & \dfrac{\partial y}{\partial R} \\ \dfrac{\partial x}{\partial \theta} & \dfrac{\partial y}{\partial \theta} \end{Vmatrix} \qquad \textbf{[5.16b]}$$

The vertical bars designate absolute value.

Let $x = R \cos \theta$ and $y = R \sin \theta$
$R = \sqrt{x^2 + y^2}$ $\tan \theta = y/x$ **[5.16c]**

where the distribution of R is the Rayleigh distribution. We obtain from Eq. [5.16b] the Jacobian transformation

$$J\left(\frac{x, y}{R, \theta}\right) = \begin{Vmatrix} \cos\theta & \sin\theta \\ -R\sin\theta & R\cos\theta \end{Vmatrix}$$
$$= R \cos^2 \theta + R \sin^2 \theta = R \qquad \textbf{[5.16d]}$$

The original and new variates are related through the vehicle of the Jacobian indicated in Eq. [5.16b]

$$q(R, \theta) = p(x, y)J = p(x, y)R \qquad \textbf{[5.16e]}$$

Hence from Eqs. [5.16a], [5.16d], and [5.16e], we obtain

$$q(R, \theta) = (\tfrac{1}{2}\pi\sigma^2)R \exp(-R^2/2\sigma^2) \qquad \textbf{[5.16f]}$$

(Note that the functions of R and θ are separable.)

The probability density function (marginal distribution) of the envelope R changes by integrating out θ, and we obtain

$$q(R) = (\tfrac{1}{2}\pi\sigma^2) \int_0^2 R \exp\left(-R^2/2\sigma^2\right) d\theta$$

$$= (R/\sigma^2) \exp\left(-R^2/2\sigma^2\right) \qquad 0 \le R < \infty \qquad \text{[5.16g]}$$

and

$$q(\theta) = (\tfrac{1}{2}\pi\sigma^2) \int_0^\infty r \exp\left(-R^2/2\sigma^2\right) dR$$

$$= \begin{cases} \tfrac{1}{2}\pi & 0 \le \theta \le 2\pi \\ 0 & \text{otherwise} \end{cases} \qquad \text{[5.16h]}$$

The distribution of θ is therefore

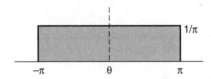

The mean value (expectation) of the phase

$$E(\theta) = \overline{\theta} = \int_{-\pi}^{\pi} \theta p(\theta)\, d\theta = 0 \qquad \text{[5.17]}$$

The mean square value is

$$E(\theta^2) = \overline{\theta^2} = \int_{-\pi}^{\pi} \theta^2 p(\theta)\, d\theta = \pi^2/3 \qquad \text{[5.18]}$$

The variance follows

$$\sigma_\theta^2 = \overline{\theta^2} - \overline{\theta}^2 = \pi^2/3 - 0 \qquad \text{[5.19]}$$

or standard deviation

$$\sigma_\theta = \pi/\sqrt{3} \qquad \text{[5.20]}$$

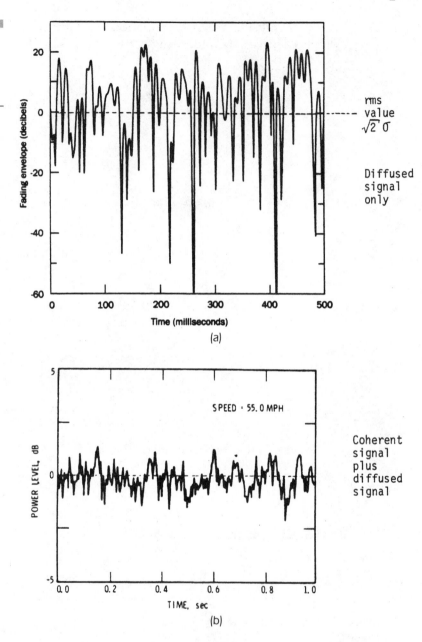

Figure 5.6
Rayleigh fading (*a*),
and coherent signal
plus Rayleigh
fading (*b*).

(a)

(b)

Data manifesting the Rayleigh distribution fading is shown in Fig. 5.6. An antenna is mounted on a vehicle moving at a speed of about 50 mph. The signal transmitted is operating at 900 MHz and there is no line of sight between the transmitter and the receiver. There is somewhat of a periodicity in the location of the fade points, which are separated by $\lambda/2$

at 900 MHz. At 900 MHz, the independent fades are about 16.7 cm (6.5 in) apart. Therefore, for vehicle movements of less than a foot, signal fluctuations can exceed 30 to 40 dB. The fading rates are also proportional to the velocity of the vehicle. It is clear that there are times of very deep fades. Even the built-in fade margins may not be able to negate these dips, and thus outage would occur for those instances. The envelope shown has a Rayleigh distribution. This is frequently referred to as a Rayleigh fading.

Additional complexity occurs since the mobile's communication back to the base station will fade at a different rate since it is transmitting at a frequency approximately 45 MHz from that received (UHF in the United States).

Note that the two previous paragraphs refer to terrestrial cellular, whose signal statistics generally manifest Rayleigh statistics. That is, the components reaching the receiver and the base station are random.

As a comparison, for operation in a more benign environment, Fig. 5.6b is a display of a signal which contains a strong, steady component coupled with a diffused multipath signal. Actually, the received signal shown is taken from experimental tests using a balloon to simulate a satellite (see Davarian). No shadowing was in evidence during these tests. The signal varies only about 3 dB peak to peak and represents a good signal. As will be discussed later, this represents a signal having Rician statistics. Note that in Fig. 5.6b the short term variation (which this is) appears to have a constant mean. But in practical environmental situations, the mean may undulate due to other intermittent fading phenomena such as lognormal shadowing. For the short term indicated, this may be considered as the local mean, but it could vary in the long term due to the changing statistics caused by shadowing or even the absence of the steady component. Indeed, there are situations where there is a mixture of Rayleigh and lognormal (diffused + shadowing) distributions. This has frequently been referred to as the Suzuki mixture (see Suzuki).

The distribution depicted in Fig. 5.6a is generally found in cellular systems where there is rarely a direct movement which would have the statistics. In cellular systems this is a difficult problem and can be alleviated somewhat with additional costs. Fortunately, in satellite-based systems where the elevation angle to the satellite is high, there will be a strong, direct line-of-sight component which will change the signal statistics to a more suitable level. That is, in addition to the Rayleigh diffused signal, there will also be a deterministic dominant signal. The signal will therefore assume Rician statistics; a discussion of the distribution follows.

5.4 Rician Distribution

In the more benign satellite-based system, there will generally be a line-of-sight component coupled with the diffused component. This will produce the Nakagami-Rician distribution. The steady component is considered a fixed vector summed with a vector whose length has a Rayleigh distribution. This is demonstrated in Fig. 5.4b.

If A is designated as the length of the fixed vector and length of the Rayleigh envelope R, the probability density function of the received envelope R is given by Rice:

$$p(v) = v \exp\left[-(v^2 + a^2)/2\right]I_0(av) \qquad 0 \le v < \infty, A \ge 0 \qquad [5.21]$$

Where I_0 equals the modified Bessel function of the first kind and zero order, whose series expansion is

$$I_0(x) = \sum_{n=0}^{\infty} \frac{x^2 n}{2^{2n} n! n!} \qquad [5.22]$$

A: *peak* value of the dominant component. The line-of-sight power is $A^2/2$.

σ^2:   average multipath power

v: R/σ Rayleigh envelope normalized to the rms multipath voltage

a: A/σ peak signal/rms multipath voltage

$I_0(x)$ is a tabulated function and converges for all x. See, for example, Jahnke and Emde (p. 226 or curve on p. 224). This is also plotted in Fig. 5.7.

When (av) becomes large, $I_0(av)$ can be replaced by its asymptotic value $(1/\sqrt{2\pi av}) \exp(av)$.* Therefore, Eq. [5.21] becomes

$$p(v) \approx (1/\sqrt{2\pi})(\sqrt{v/a}) \exp\left[-(v-a)^2/2\right] \qquad [5.23]$$

The distribution assumes a normal law. Note that the average value of the curve is A/σ and the standard deviation is σ. If we assume that the multipath is nil (and we assume that the receiver thermal noise is likewise) and the signal is sinusoidal, the output becomes $u_0(v - A)$, or impulsive.

A plot of the Nakagami-Rician distribution is shown plotted in Fig. 5.8. From Eq. [5.15] and the figure, for $A' = 0$, the distribution reverts to the

---

* $I_0(x) \approx (1/\sqrt{2\pi x}) \exp\left[1 + (1^2/1!8x) + (1^23^2/2!(8x)^2 + \cdots\right]$.

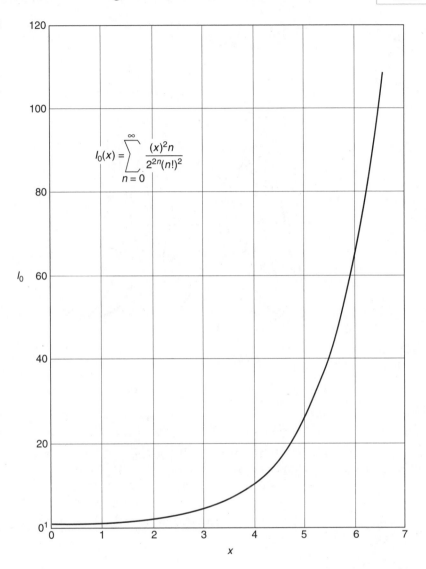

Figure 5.7
Plot of the Bessel function of the first kind of order zero.

$$I_0(x) = \sum_{n=0}^{\infty} \frac{(x)^2 n}{2^{2n}(n!)^2}$$

Rayleigh distribution. For large values of A', the curves approach normality, and the mean value approaches A' with standard deviation equal to σ. This would occur for values of greater than 5 (approximately). It is also noticed that as A' increases, the peaks of the curves approach $1/\sqrt{2\pi}$ as an asymptote.

The Nakagami-Rician distribution originally stemmed from S. O. Rice's classical paper in the *Bell System Technical Journal*. Rice's analysis was centered on the envelope of a sine wave plus additive band-pass gaussian

■■■ ■■■ ■■■ ■■■

Figure 5.8
March of the Rician
distribution for para-
metric values of
$A' = A/\sigma$.

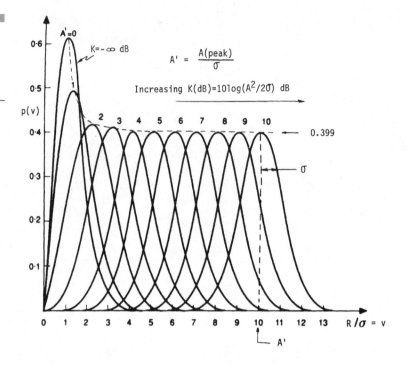

• It does not take much of a march to get from the Rayleigh
  distribution (A'=0) to the Gaussian distribution (A' ≈ 5).

noise.* The reader familiar with Rice's paper may have observed that he plotted the density functions using $a = A'/\sigma$ being parametric and has been referred to as the signal-to-noise ratio. Actually, he plotted *peak* voltage over rms value σ. To be strictly correct, rms voltage is peak$/\sqrt{2} = A/\sqrt{2}$ for a sinusoid. Therefore, the $a = A/\sigma$ should be $P/\sqrt{2}\sigma$, or for power $S/N = A^2/2\sigma^2$. Therefore, a is larger than the rms signal-to-noise ratio by the factor $\sqrt{2}$.

5.5 Moments of the Rice Distribution

In Rice, the general equation for the first-order moments (about zero) of the Rice distribution is given as

---

*In the analogy between passband noise and diffused multipath, the diffused compo-
nent acts like a noise term. Nakagami may have realized this connection and thus the designation Nakagami-Rician distribution.

$$\overline{e^n} = (2\sigma^2)^{n/2}\Gamma[(n/2) + 1]\exp(-A^2/2\sigma^2)\,_1F_1[(n/2) + 1; 1; (A^2/2\sigma^2)]$$

$$= (2\sigma^2)^{n/2}\Gamma[(n/2) + 1]\,_1F_1[(-n/2); 1; (-A^2/2\sigma^2)] \qquad \textbf{[5.24]}$$

where σ = average noise power

    $_1F_1$ = confluent hypergeometric function

    A = peak signal voltage incident on detector

  $A^2/2\sigma^2$ = rms power signal to noise ratio

If the deterministic signal A is not present, the $(S/N) = 0$, and the first moment is

$$\overline{e} = (2\sigma^2)^{1/2}\Gamma(3/2) = \sqrt{\pi/2}\,\sigma = 1.25\sigma \cdots \qquad \textbf{[5.25]}$$

Since

$$_1F_1[(-\tfrac{1}{2}); 1; 0] = 1 \qquad \textbf{[5.26]}$$

The second moment is

$$\overline{e^2} = 2\sigma^2\Gamma(2)[1 + (A^2/2\sigma^2)] = 2\sigma^2 \qquad \textbf{[5.27]}$$

The second centroidal moment is

$$\sigma'^2 = E(e - \overline{e})^2 = \overline{(e - \overline{e})^2} = \int_0^{\infty} (e - \overline{e})^2 p(e)\,de \qquad \textbf{[5.28]}$$

where $p(e) = (e/\sigma^2)\exp(-e^2/2\sigma^2)$, which is the Rayleigh distribution. Continuing, we obtain

$$\sigma'^2 = \overline{e^2} - (\overline{e})^2 = \text{mean square} + \text{square of the mean} \qquad \textbf{[5.29]}$$

and from Eqs. [5.25] and [5.27]

$$\sigma'^2 = \sigma^2[2 - (\pi/2)]$$

$$= 0.430\sigma^2$$

$$\therefore \; \sigma' = 0.655\sigma \qquad \textbf{[5.30]}$$

For a finite S/N ratio $(A \neq 0)$, the first moment is

$$\overline{e} = \sqrt{2}\sigma\Gamma(\tfrac{3}{2})\,_1F_1[(-\tfrac{1}{2}); 1; (-A^2/2\sigma^2)]$$

or alternately, if Bessel functions are more readily available

$$\overline{e} = \sqrt{2}\sigma\Gamma(\tfrac{3}{2})\exp(-A^2/4\sigma^2)\{[1 + (A^2/2\sigma^2)]I_0(A^2/2\sigma^2) + (A^2/2\sigma^2)I_1(A^2/4\sigma^2)\} \qquad \textbf{[5.31]}$$

where I_0 = Bessel function of zero order and imaginary argument
I_1 = Bessel function of first order and imaginary argument

$$\Gamma(\tfrac{3}{2}) = \sqrt{\pi}/2 \text{ gamma function} \qquad [5.32]$$

A trajectory of the mean, as a function of rms power S/N ratio, can be found using the plot of the hypergeometric function shown in Fig. 5.9.

The excursion of the mean, with increased rms power signal-to-noise ratio, is shown plotted in Fig. 5.10. The ticks along the abscissa scale indi-

Figure 5.9
Confluent hypergeo-
metric function.

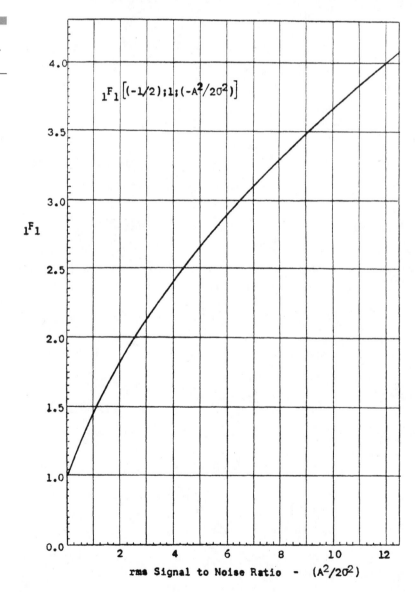

$$_1F_1\left[(-1/2);1;(-A^2/2\sigma^2)\right]$$

$_1F_1$

rms Signal to Noise Ratio — $(A^2/2\sigma^2)$

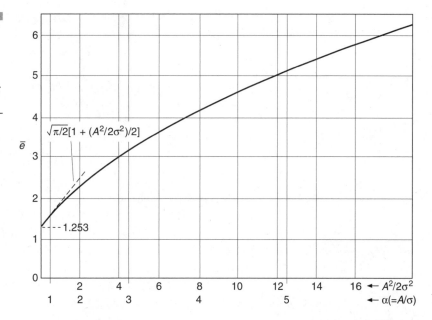

Figure 5.10
Trajectory of the mean of the Rice distribution with increasing rms power signal-to-noise ratio.

cate discrete values of A $(=A/\sigma)$ which give the same mean value. Even though not explicitly shown, for increasing values of a, the value of the mean is $\bar{e} \to A$. This can be shown by using the asymptotic expansion of $_1F_1(-)$; which is convenient to use for large values of $(A^2/2\sigma^2)$. This expression is:

$$_1F_1(a;\,\beta;\,-z) \approx \frac{\Gamma(\beta)}{\Gamma(\beta-a)}\,z^{-a}[1 + a(1 + a - \beta)/z$$
$$+ a(a+1)(1 + a - \beta)(2 + a + \beta)/z^2 2! + \cdots] \qquad \textbf{[5.33]}$$

From Eqs. [5.31] and [5.33] we obtain

$$\bar{e} \approx 2\sigma\Gamma(\tfrac{3}{2})[(2A/\sqrt{2\pi}\sigma)(1 + [2\sigma^{\overset{\approx 0}{\cancel{2/4A^2}}}])]$$
$$\approx A \qquad \textbf{[5.34]}$$

We also notice if $_1F_1(-)$ is put in the form of a power series expansion, the value of $\bar{e}$ becomes

$$\bar{e} = \sqrt{2}\sigma\Gamma(\tfrac{3}{2})[1 + (\tfrac{1}{2})(A^2/2\sigma^2) + (\tfrac{1}{16})(A^2/2\sigma^2)^2 + (\tfrac{1}{96})(A^2/2\sigma^2)^3 + \cdots] \qquad \textbf{[5.35]}$$

and for small values of signal power, or coherent received power, we obtain

$$\bar{e} \approx \sqrt{2}\sigma\Gamma(\tfrac{3}{2})[1 + (\tfrac{1}{2})(A^2/2\sigma^2)] \qquad \textbf{[5.36]}$$

The second moment for a finite signal-to-noise ratio is given as

$$\overline{e^2} = 2\sigma^2 \Gamma(2)[1 + (A^2/2\sigma^2)] = 2\sigma^2 + A^2 \qquad [5.37]$$

The second centroidal moment (variance) for the blossomed Rice distribution follows from Eqs. [5.34] and [5.37]

$$\gamma^2 = \overline{e^2} - (\overline{e})^2$$
$$= 2\sigma^2\{[1 = (A^2/2\sigma^2) - (\pi/4)_1F_1(-1/2); 1; (-A^2/2\sigma^2)]^2\} \qquad [5.38]$$

This relationship is shown plotted with respect to rms power signal-to-noise ratio and also normalized with respect to σ in Fig. 5.11. Also shown is the plot for normalized γ. Similar to Fig. 5.8, the ticks give the value of a ($=A/2$) corresponding to a particular value of $A^2/2\sigma^2$. The standard deviation of the output assumes the value of the input noise voltage. We can show this simply by noting for large signal-to-noise ratios from Eqs. [5.23] and [5.25], we obtain

$$\gamma = \overline{e^2} - (\overline{e})^2$$
$$= 2\sigma^2[1 + A^2/2\sigma^2] - A^2$$
$$= 2\sigma^2 = \text{peak value } (\sqrt{2}\sigma)^2, \text{ the rms value is } \sigma^2 \qquad [5.39]$$

Figure 5.11
Excursion of the normalized standard deviation and second centroidal moment of the Rician distribution with increasing rms power *S/N* ratio.

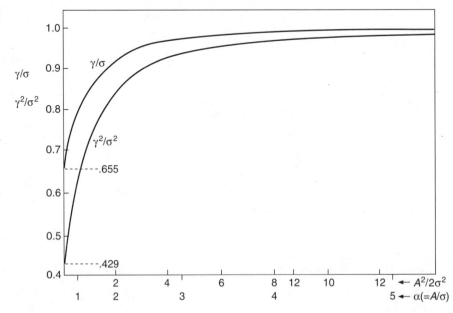

The probability density function of the phase distribution for the Rician distribution is given in Bennett (see References). (Rice uses the symbol θ for phase in his classical paper on noise.)

$$p(\phi) = \exp(-A^2/2\sigma^2)$$
$$+ (A \cos \phi/2\sigma\sqrt{2\pi})[1 + \mathrm{erf}(A \cos \phi/\sqrt{2\sigma})] \exp(-A^2 \sin^2 \phi/2\sigma^2) \quad [\mathbf{5.40}]$$

and

$$\sigma_\phi = \pi/\sqrt{3}[1 - (\sqrt{9/2\pi^3})(v/\sigma)] \quad [\mathbf{5.41}]$$

Note that when $A = 0$, the equation reduces to

$$p(\phi) = 1/2\pi \qquad 0 \leq \phi \leq |\pi| \quad [\mathbf{5.42}]$$

and

$$\sigma_\phi = \pi/\sqrt{3}$$

For large S/N, the probability density is approximately gaussian and is given by

$$p(\phi) = (1/\sqrt{2\pi})(v/\sigma) \exp(-v^2\phi^2/2\sigma^2) \quad \text{and} \quad \sigma_\phi = \sigma/v \quad [\mathbf{5.43}]$$

For $v/\sigma \to \infty$, the probability density function becomes impulsive and is centered on the phase of the signal (no Doppler considered). The phase functions are displayed in Fig. 5.12.

If we normalize the ratio dividing by the rms diffused power, the line-of-sight power is given by

$$P_{\mathrm{LOS}} = A^2/2 \quad [\mathbf{5.44}]$$

And the diffused or multipath power is

$$P_d = \sigma^2 \quad [\mathbf{5.45}]$$

The total power at the receiver is the power in the direct component plus the power in the difference component. It is shown by

$$P_T = A^2/2 + \sigma_d^2 \quad [\mathbf{5.46}]$$

The ratio of these two are frequently referred to as the K factor, where $(a^2/2)$ is the power in the steady component, and σ^2 is the power in the diffused component. Thus,

$$K = P_{\mathrm{LOS}}/P_d = A^2/2\sigma^2 \quad [\mathbf{5.47}]$$

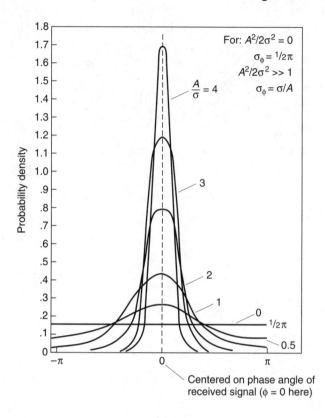

Figure 5.12
Phase fluctuation for sine wave plus gaussian noise.

also given as

$$K_{db} = 10 \log (P_{LOS}/P_d) = 10 \log (A^2/2\sigma^2) \qquad \text{[5.48]}$$

In fact, the Rician distribution is occasionally given in terms of the K factor. Therefore, the Rician distribution can be plotted in terms of K rather than a. We therefore have

$$p(r) = \frac{2rK}{a^2} \exp\left[-\frac{K}{a^2}(r^2 + \sigma^2)\right] I_0\left[\frac{2rK}{a}\right] \qquad \text{[5.49]}$$

(If defined in dB, replace K by $10^{\kappa/10}$.) When $K \to 0$, the distribution tends to Rayleigh. When $K \gg 1$, the distribution tends to be a gaussian distribution in which the mean is equal to a (see Fig. 5.8). The gaussian profile can be shown to be true by plotting the Rician distribution on arithmetic probability paper. The $A = 5$ curve will plot practically linear. For $A < 5$, the curves will be bowed. This is depicted in Fig. 5.13. It is interesting to note on arithmetic probability paper that the curves maintain

Figure 5.13

Distribution function of envelope R of $I(t) = \cos(\omega t) + \sigma$, modified from S. O. Rice, "Math. Anal. of Random Noise" (see N. Wax edition, p. 241).

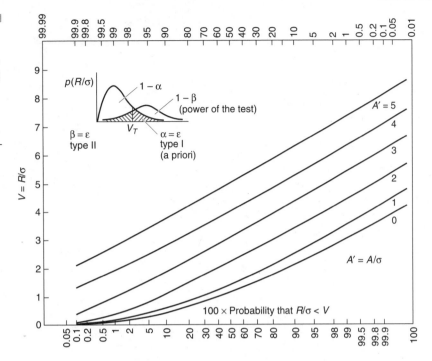

their linear slope and will merely translate as the signal-to-noise ratio (or diffused signal) increases. The constant slope indicates that there is no change in the variance for increasingly large S/N ratios. A changing variance (with constant signal) would rotate the curve and for $\sigma \to 0$, the slope is horizontal with rotation about the mean or a. As noted, the mean is independent of the variance or sigma.

If these curves are plotted on Rayleigh probability paper, it is clear that only in the case where $a = 0$ will the curve be linear. See Fig. 5.14. Curves for $a > 0$ are bowed. Also, because of skewness of the density function, the 50 percent probability point is closer to the skewed direction. As the signal-to-noise ratio becomes larger, the curves become more nonlinear indicating a greater departure from the Rayleigh law. It may also be observed that changing the value of σ, for $a = 0$, the Rayleigh curve is translated upward if σ increases and is not a rotation of the curve.

Typical values of K are about 10 dB (with residual fading), but, of course, this depends on the strength of the steady sector. Large values of K (fading) suggest a large dominant component which generally occurs in satellite applications. With heavy fading, K is small, and the signal assumes Rayleigh statistics. The latter certainly prevails in terrestrial cellular systems, and the system designer is ever combatting this diffused

Figure 5.14

Rayleigh probability
paper representation
of the Rice single look
marginal amplitude
distribution for dis-
crete signal-to-noise
ratios ≤ 5.

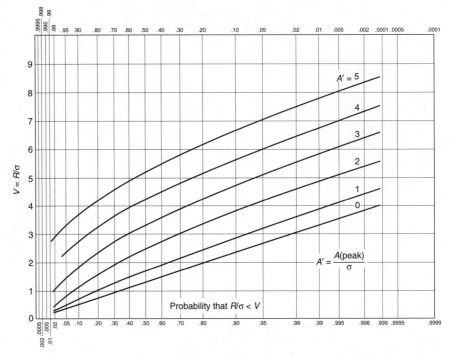

fade to increase system availability (less outage) or reliability (less atten-uation).

In digital systems, fading has a deleterious effect on the desired bit error rate (BER). For example, in white gaussian noise, a BER = 10^{-3} requires an $E_b/N_0 = 6.9$ dB for BPSK.* In a fading situation, E_b/N_0 required may be 10 to 20 dB higher to realize the same BER. Recourses one has are: (1) to increase the fade margin (at cost), (2) to use error correction coding (FEC), and (3) to use RAKE receivers to benefit from the multipath signals.

Some simulations performed by Davarian on the effects of fading are shown in Fig. 5.15. Figure 5.15a shows results in the presence of diffused or Rayleigh fading only. The ideal and static measurements (accounts for implementation loss) are made in the absence of any fading and results from fading. In Fig. 5.15a, results are shown for three Doppler conditions. There is an appreciable degradation in BER resulting from Rayleigh fading.

Figure 5.15b shows Rician fading. The degradation is not quite as severe because of the presence of a strong, dominant signal component. Note the value of $K = 10$, where the dominant component is 10 dB stronger than the diffused component.

---

*BPSK is the binary phase shift keying.

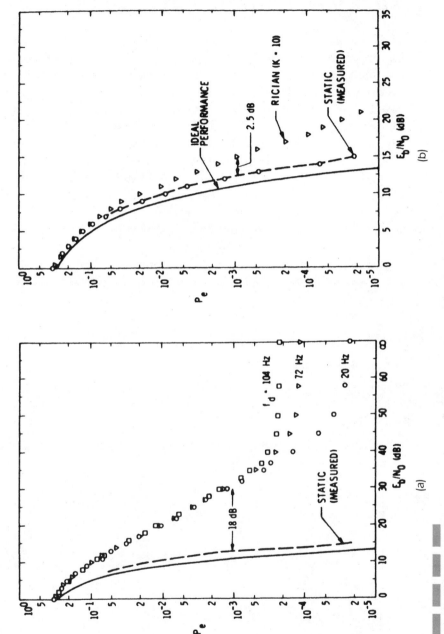

Figure 5.15 (a) The error performance of the noncoherent FSK receiver in the presence of Rayleigh fading, and (b) the error performance of noncoherent FSK receiver in the presence of Rician fading.

133

5.6 Lognormal Distribution

The propagation scenario sketch depicted in Fig. 5.1 shows signal shadowing impairment which also degrades performance. This is shadowing of the direct signal by foliage and other obstacles which can attenuate and distort the signal. This can be an especially severe problem for satellite-based systems where the elevation angles are small at the higher latitudes for geostationary satellites or any latitudes for low earth orbit satellites. The signal statistics change depending on whether there is shadowing or not. The unshadowed cases have been discussed in previous paragraphs. For line-of-sight signals which are shadowed by foliage, hills, or other obstacles, experimental data indicates that the amplitude of the receiver signal assumes lognormal statistics. That is, the totality of points falls in a straight line when plotted on a graph where the abscissa scale is arithmetic probability and the ordinate is in signal dB values.

In the presence of shadowing, the amplitude of the direct wave (shadowed) assumes a lognormal distribution with a probability density function (see Bocthias).

$$p(v) = (1/\sqrt{2\pi}\sigma v) \exp[-(\ln v - m)^2/2\sigma^2] \qquad 0 \le v \le \infty \qquad [5.50]$$

A plot of this function for different values of σ^2 is shown in Fig. 5.16. The cumulative distribution function is

$$P(v) = (\tfrac{1}{2})[1 + \mathrm{erf}\,(\ln v - m/\sigma\sqrt{2})] \qquad\qquad [5.51]$$

where v is the amplitude, and m and σ are the mean and standard deviation not of the variable v, but the logarithm of this variable (m is mean of

Figure 5.16
Probability density function for lognormal distribution for difference values of the variance.

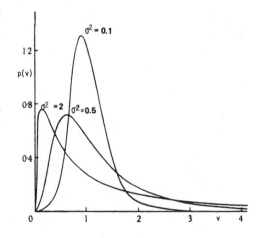

log v, σ is standard deviation of log v). The signal voltage (amplitude) is lognormally distributed, and the signal power in dB is normally distributed.

The various attributes of this distribution are as follows.

- Most probable value (σ) = exp $(m - \sigma^2)$ (mode)
- Median value = exp (m)
- Mean value = exp $(m + [\sigma^2/2])$
- Root mean square value = exp $(m + v^2)$
- Standard deviation = exp $(m + \sigma^2/2)\sqrt{\exp(\sigma^2) - 1}$

Measurements through foliage made in Canada (see Butterworth) and elsewhere have clearly established that a discrete signal propagation through foliage produces data in which the underlying distribution is lognormal.

The results of some tests performed at 870 MHz through foliage in Canada (see Butterworth) are shown in Fig. 5.17. These were static tests with a transmitter on a tower as illustrated. The totality of measurement results in dB on gaussian probability paper shows the linear relationship. The mean value of this attenuation is 7 dB. Any curve which is parallel with these data will have a different mean, but the standard deviation will remain the same. If the curve rotates, there will be a change in the standard deviation.

In a dynamic situation where a receiving antenna is mounted on a moving vehicle, one can expect the signal statistics to change. En route, one can observe periods when the signal is in the clear with a very large K factor, a signal mixed with diffuse fading, one that is shadowed, and even a mixture of shadowed plus diffused. Reflected spectral energy is also possible, but this is generally much less than the direct energy. Down 10 to 15 dB may be typical but clearly depends on the terrain composition. The reflection coefficients for several terrain compositions are shown in Fig. 5.18. (See Sue and Park.) In Fig. 5.18a, observe that for vertical polarization the reflection coefficient drops from low elevation angles to zero and the phase shift is around 180°. Below the Brewster angle there is no phase shift. Above the Brewster angle, the reflection coefficient increases but does not reach the zero elevation angle level. At the Brewster angle an interesting phenomenon occurs: there is a phase shift of around 180°. If the radiation received is circularly polarized (this will be the polarization from the satellite because of its insensitivity to Faraday rotation), there will be a sense of polarization shift. For example, a right-hand circular polarized wave will have its sense of rotation reversed to left-hand circular polarization. Therefore, an antenna designed to accept right-hand

Figure 5.17
Cumulative distribu-
tion of foliage
attenuation.

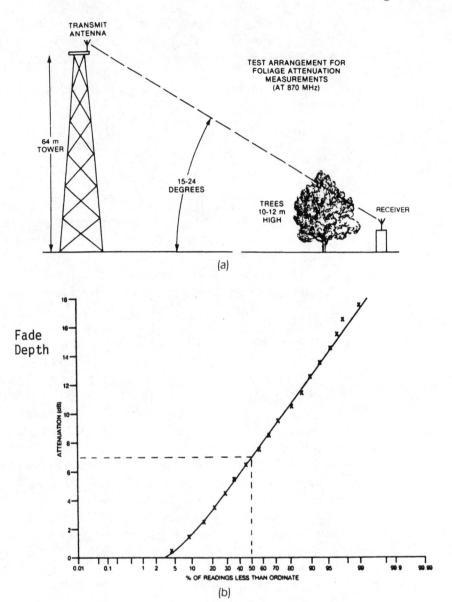

TRANSMIT
ANTENNA

TEST ARRANGEMENT FOR
FOLIAGE ATTENUATION
MEASUREMENTS
(AT 870 MHz)

64 m
TOWER

15-24
DEGREES

TREES
10-12 m
HIGH

RECEIVER

(a)

Fade
Depth

ATTENUATION (dB)

% OF READINGS LESS THAN ORDINATE

(b)

circular polarization will reject the left-hand circular polarization.
Clearly, this desirable feature helps to eliminate multipath.

The horizontally polarized wave (see Fig. 5.18*b*) appears to behave less vio-
lently than the vertically polarized wave. The reflection coefficient drops
monotonically but more slowly than the vertical polarized signal. However,
at zero elevation angle, both are comparable. At zenith (90°), they are again

Figure 5.18
Plane-earth reflection
coefficient for 15 GHz
(but applicable at
1 GHz).

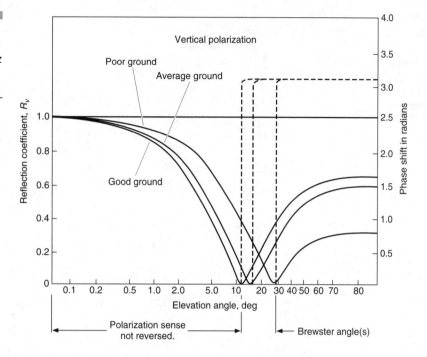

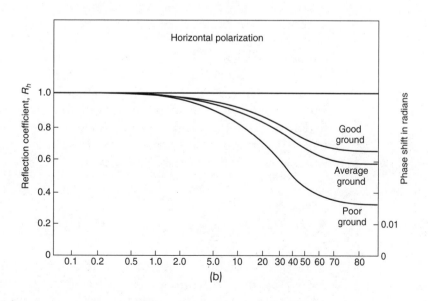

Figure 5.18
Plane-earth reflection
coefficient for 15 GHz
(but applicable at
1 GHz).

- Earth reflections is a function of earth conductivity, dielectric constant and frequency.

(a)

(b)

Figure 5.19
Comparison of measurements made at roughly 800 MHz by independent researchers.

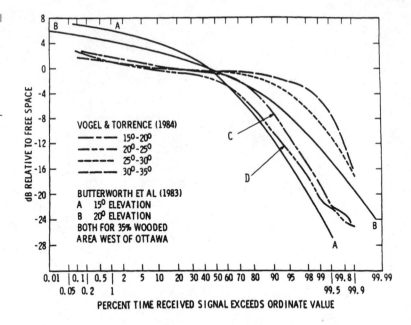

- The inconsistency between "C" (lower elevation angle) and "D" (higher elevation angle) is due to denser foliage penetration in the "D" measurements. Clearly, one would expect less foliage attenuation at higher elevation angles.

comparable but at reduced values. The phase shift of the reflected wave is always out of phase with the incident wave, that is, about 180°.

Some of the data gleaned for the above conditions from various investigations are shown in Fig. 5.19 (see Sue and Park). This is shown for various elevation angles. The curves are divided into essentially two parts. The relatively flat position to the left of the knee has basically a Rician distribution which includes a dominant component plus diffused multipath, and no shadowing exists. To the right of the knees, shadowing is manifested and has a lognormal distribution. For lower elevation angles (more foliage penetration), the knees of the curves shift further to the left, there being less direct line-of-sight operation.

5.7 Nakagami-*m* Distribution

Another distribution frequently encountered in propagation studies is the Nakagami-*m* distribution. This is a more general distribution and relaxes restrictions on the amplitude and phase of the waves. For example, for the Rayleigh distribution the received signal is the sum of vectors of comparable magnitude and independent, uniformly distributed phases. In the Nakagami-m distribution there is no assumption as to the statistics of the amplitudes and phases other than being random. The probability density function can be written as follows, with two parameters Ω and m.

$$p(r) = \frac{2m^m r^{2m-1} \exp\left[(-m/\Omega)r^2\right]}{\Gamma(m)\Omega^m} \qquad m \geq 1/2, \; r \geq 0 \qquad [5.52]$$

where Ω = time average power (mean square) of the received signal
    m = shape factor (mean square)

The various distributions alluded to previously (Rayleigh, Rician, and lognormal) can be derived from the Nakagami-m by a proper choice of the parameters. For example, for $m = 1$, the Rayleigh probability density function is obtained. For $m > \infty$, there is no fading. For $1 < m < \infty$, the Rician and lognormal distributions are realized. A plot of this cumulative distribution for various values of m ($\Omega = \overline{r^2} = 1$) is shown in Fig. 5.20a (see Bocthias). Note the Nakagami-m distribution becomes Rayleigh for $m = 1$. Experimental data (see Jahnke and Emde and Bennett) in Fig. 5.20b shows that empirical data straddles the Rayleigh distribution. For $u > 1$, the distribution changes and resides in the domain of Rician and/or lognormal. With increasing m, the curve for Nakagami-m becomes more gaussian in shape which the reader recalls in which the Rician distribution also becomes gaussian with increasing strength of the line-of-sight signal.

A plot of the Nakagami-m probability density function, for various values of m ($\Omega = 1$), is shown in Fig. 5.21. As indicated previously, this is a generalized function making no assumptions as to the statistics of the signal. For $m = 1$ the Rayleigh density function evolves. For $m = 8$, this approximates the Rician distribution for a large fixed vector with a small fading term. Clearly for $m \to \infty$, there is increasingly less fading. Since the time average power ($\Omega = 1 = \overline{r^2}$) or mean square is constant or normalized as m becomes larger and larger, the bell-shaped curve will become more impulsive and taller at $r = 1$, and standard deviation becomes smaller since total power is constant. This is intuitively satisfying since the signal is becoming more and more deterministic and less statistical, and the fading effects are becoming vanishingly smaller.

5.8 Summary

Signals propagation in wireless communications, be it terrestrial or satellite-based, are not deterministic and must be analyzed on a statistical basis.

Figure 5.20
The Nakagami-*m* dis-
tribution and experi-
mental data.

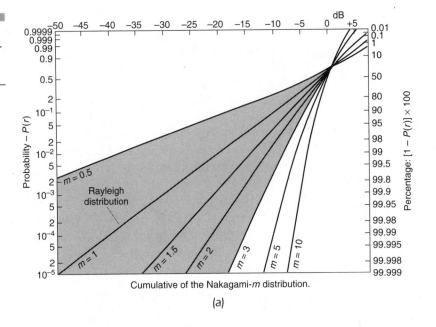

Cumulative of the Nakagami-*m* distribution.

(a)

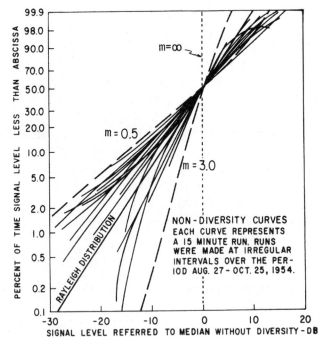

Empirical distributions (solid lines, from [16]) and two
Nakagami distributions (dashed lines).

(b)

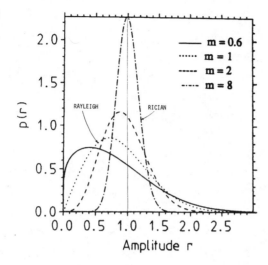

Figure 5.21
The march of the Nakagami-*m* distribution.

The signal statistics can vary from a rarely observed purely benign signal whose magnitude is deterministically defined to complicated structures which are affected by externals and must be defined in statistical terms. It is therefore essential that channel characterization be fully understood to explain their vagaries. Analytic studies coupled with field measurements are necessary. Also for satellites in LEO/MEO even if the earth station is relatively fixed.

In previous material, we have defined some of the channel perturbations and the resulting signal formats. The anomalies are principally multipath, blockage, and shadowing, and for mobile units, Doppler-induced dynamics must be considered.

It is hoped that the reader can use some of these concepts to improve the performance of wireless systems, which reflects in greater link availability or reduced outage, and ultimately, for commercial applications which are the bottom line.

References

Beckmann, P., and A. Spizzichino, *The Scattering of Electromagnetic Waves from Rough Surfaces*, Pergamon Press Books, New York, 1963.

Bennett, W. R., "Methods of Solving Noise Problems," *Proc. IEEE*, May 1956.

Bocthias, L., "Radio Wave Propagation," McGraw-Hill Co., New York, 1987.

Bulington, K., et al., "Results of Propagation Tests at 505 MHz and 4090 MHz on Beyond-Horizon Paths," *Proc. IRE,* October 1955.

Burington, R., and D. May, *Handbook of Probability & Statistics,* Handbook Publishers, Inc., Sandusky, Ohio, 1953.

Butterworth, J. S., "Propagation Measurements for Land-Mobile Satellite Devices in the 800 MHz Band," CRC Tech. Note 724, Gov. of Canada, August 1984.

Charash, U., "Reception Through Nakagami Fading Multipath Channels With Random Delays," *IEEE Trans. on Communications,* April 1979.

Davarian, F., "Channel Simulation to Facilitate Mobile Satellite Communications Research," *IEEE Trans. on Communications,* January 1987.

Flock, W. L., "Propagation Effects on Satellite Systems at Frequencies Below 10 GHz," NASA Ref. Publ. 1108(02), December 1987.

Hata, M., "Empirical Formula for Propagation Loss in Land Mobile Radio Devices," *IEEE Trans. Vehicular Technology,* vol. VT-29, August 1980.

Hoel, P. G., "Introduction to Mathematical Statistics," 2d ed., J. Wiley & Sons, New York, 1958.

ITU CCIR Rpt. 263-7, annex to vol. vi, "Ionospheric Effects Upon Earth-Space Propagation," Düsseldorf, 1990.

ITU CCIR Rpt. 527-1 "Electrical Characteristics of the Surface of the Earth," vol. v, 1982.

Jahnke, E., and F. Emde, *Table of Functions,* Dover Pub., 4th ed., 1945.

Middleton, D., and V. Johnson, "A Tabulation of Selected Confluent Hypergeometric Functions," T.R. #140, Cruft Laboratory, Harvard University, January 1952.

Miller, K. S., *Engineering Mathematics,* Rinehart & Co., New York, 1957.

Pattan, B., "The Graphics of the Rayleigh Distribution," Sylvania (GTE) System Engineering Laboratory, Internal Report, 1964.

Pattan, B., "Second Centroidal Moments of the Rice Distribution Through the Vehicle of the Torsion Pendulum," Sylvania (GTE) Engineering Laboratory, Internal Report, 1965.

Pattan, B., "Lectures on Random Noise Theory," course given at Lockheed Inc., October 1967.

Pattan, B., "Fifteen Useful Curves Stemming from the Q Function" (Marcum), Computer Sciences Corp., Internal Report, 1978.

Rice, S. O., "Mathematical Analysis of Random Noise," *BSTJ* July 1944/*BSTJ* Jan. 1945. Also published in "Noise and Stochastic Processes," N. Wax (ed.), 1954.

Schwartz, M., et al., *Communications Systems and Techniques,* McGraw-Hill Co., New York, 1966.

Slack, M., "The Probability Distributions of Sinusoidal Oscillations Combined in Random Phase," *P-IEE,* pt. III, 1946.

Sue, M. K., and Y. H. Park, "Second Generation Mobile Satellite System," JPL Pub. 85-88, June 1985.

Suzuki, H., "A Statistical Model for Urban Radio Propagation," *IEEE Trans. on Communications,* no. 7, July 1977.

Vogel, W., and E. Smith, "Propagation Considerations in Land Mobile Satellite Transmission," *Microwave Journal,* October 1985.

Watson, G. N., *Theory of Bessel Functions,* 2d ed., Cambridge University Press, Cambridge, 1944.

CHAPTER 6

Modulations Used in Cellular Digital Communications

6.1 Introductory Comments

The burgeoning advances made in telecommunications have created an explosive use of the spectrum, creating a paucity of allocated spectrum to serve all of these needs. Over the last decade or so, schemes have been devised to make better use of the existing spectrum. Some of the earlier attempts have been via trunking and the use of single sideband modulation.

Trunking is basically the automatic sharing of communication channels by a group of users. Prior to that, dedicated channels were used and thus limited the capacity of systems. An additional benefit was a relief in blocking. However, there are applications where dedicated channels are required, and these are still in use.

Single sideband generation is extensively used, and the FCC has played a significant role in promoting its development. With amplitude-companded single sideband, one can jam more channels into a channel which previously used FM. For example, six channels less than 5 kHz wide can replace a single FM channel of 30 kHz wide. Additionally, one can insert an SSB signal into a guard band between two FM channels.

Cellular structuring concepts are clearly another way to realize an improvement in spectral efficiency. In lieu of using a single transmitter to serve a single large area or cell, the service area is broken down into a smaller cluster of cells where each cell is assigned a group of channels. In the original one-cell concept, the allocated band was broken up into channels (e.g., 30 kHz wide for FM) and each channel or that sliver of the spectrum cannot be shared. The cellular radio concept increased the capacity of the system by the development of frequency reuse.

Some more recent advances made in the improvement of spectral efficiency have been in the use of digital communications and, in particular, in the area of source coding and channel coding and modulation.

Considerable research has been going on in recent years in the area of source coding, that is, reducing the number of bits required to represent speech. However, there may be some caveats here. A low-bit coder may indeed provide greater capacity, but the speech quality may suffer. Several strata of speech quality have been identified: (1) toll quality, (2) communication quality, and (3) synthetic quality. The first is generally provided by bit rates in the range of 64 to 32 kbps. The second is down to about 8 kbps with research pushing 4.8 kbps. Communication quality has frequently been referred to as near-toll quality performance. Bit rates below, say, about 5 kbps may not have a naturalness about it. These ranges may be more suitable for military applications, but in the commercial sector may not see light of day in telephony. An example of the latter is linear predictive coding (LPC) at 2.4 kbps. The North American IS-54 standard employs an 8-kbps coder. The European GSM system uses a 13-kbps coder. Higher bit-rate decoders generally give better performance, but clearly at the expense of greater bandwidth. However, research is in pursuit of near-toll quality speech with a 4.8-kbps vocoder. JPL and its contractors are very active in this area. Note that ACSB analog voice can now occupy less than 5 kbps bandwidth and may be replaced by digital at 4.8 kbps. Realizing commercially available quality voice at lower bit rates may not be feasible. The bottom line is what quality speech will be acceptable and at what price.

There are basically three areas of signal efficiency which are bandied around in communication circles: (1) bandwidth, (2) spectral, and (3) communication efficiency. Bandwidth efficiency defines the number of bits one can realize per hertz of bandwidth. This is information density in the shannon sense. In spectral efficiency, one attempts to adequately represent a signal with the minimum amount of bandwidth and with little spectral spillover (which can cause interference to adjacent channel(s) or adjacent symbol interference). Bandwidth in this context may mean different values to researchers. How much bandwidth is required to adequately represent a signal? Is it 99 percent, 90 percent, or 3-dB bandwidth of the power spectral density? The more the truncation of the spectral lobes, the more energy is discarded. We therefore have a trade-off. In communication efficiency, attempts are made to use the minimum power to realize the desired bit error rate (BER). This bit error rate results from the use of coherent or noncoherent detection in the receiver and the use of error-correction schemes to realize coding gain.

In the past 30 years, we have seen an evolution in communications. Some of the advancements have been breakthroughs while others were embellishments or refinements. One advancement which has become quite evident is the trend from analog to digital modulation. Clearly, two main reasons for this migration have been the advent of computers and the design of circuits using VLSI. Other reasons include coding to improve the communication efficiency and source coding to increase the capacity.

New modulation techniques are spectrally efficient permitting closer channel spacing, are constant envelope signals, can be amplified by efficient near-saturation power amplifiers, and can reduce spectral spillover which would produce adjacent channel interference (ACI). Concomitant with these modulations, it is also desirable to use the minimum energy to realize adequate bit error rates.

6.2 PSK, QPSK, and OQPSK

The classical digital modulation schemes which are currently in use, such as BPSK and QPSK, have performed adequately to date. However, with the paucity of available spectrum and the advent of new services, their spectral occupancy has left much to be desired. The signal bits (which determine the spectral distribution) manifest spectral sidelobes which fall at best at a $1/f^2$ rate. These lobes cause intersymbol and interchannel interference. One reason for these sidelobes is the abrupt phase transitions made at the intersymbols. Another is the sharp corners on the pulses. These phase changes can be $0°$, $\pm90°$, or $\pm180°$.

One improvement in these phase changes has been the use of offset QPSK (sometimes called staggered QPSK), where the bit transitions result in phase changes of $0°$, or $\pm90°$, excluding the phase changes of $\pm180°$ of QPSK. Offsetting of the bits is shown in Fig. 6.1, in comparison to QPSK. Offsetting of bits does not change the power spectral density, and the spectrum of OQPSK is the same as QPSK, as shown in Fig. 6.2. Note that the spectral falloff is $1/f^2$, the same as BPSK. These are also constant envelope signals (prior to filtering), and in nonlinear channels the signals are not affected since the information is in the phase. However, if QPSK is band-limited to truncate the spectral spillover, the signal is *no longer* a constant envelope. Limiting will return the signal to constant envelope but will also restore the spectral sidelobes of QPSK. OQPSK, when band-limited, does not lose its constant envelope property even though there is some droop at the $0°$ and $\pm90°$ phase transitions. Any further processing (like limiting) which may occur when an amplifier is operating near saturation for maximum efficiency will restore the sidelobes in QPSK, but not in OQPSK. It is the avoidance of the $180°$ phase transitions which makes possible reduced out-of-band interference and contributes to the spectral efficiency.

Even though the OQPSK manifests an improvement in phase transitions and reduced out-of-band interference, it still manifests some interference since discrete discontinuous phase transitions ($0°$, $\pm90°$) do occur. It naturally follows that, if one can obviate impulsive phase transitions, this will help to reduce the sidelobes. Some relatively recent research has brought forth continuous phase modulation techniques with signals hav-

Figure 6.1
QPSK and OQPSK
modulation
generation.

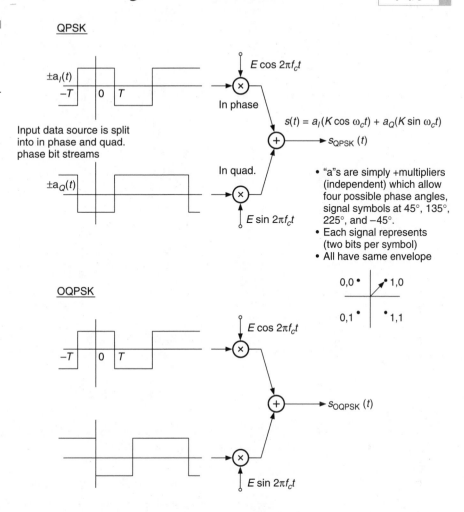

QPSK

$\pm a_I(t)$

$-T$ 0 T

In phase

$E \cos 2\pi f_c t$

Input data source is split
into in phase and quad.
phase bit streams

$\pm a_Q(t)$

In quad.

$E \sin 2\pi f_c t$

$s(t) = a_I(K \cos \omega_c t) + a_Q(K \sin \omega_c t)$

$s_{QPSK}(t)$

- "a"s are simply +multipliers
  (independent) which allow
  four possible phase angles,
  signal symbols at 45°, 135°,
  225°, and −45°.
- Each signal represents
  (two bits per symbol)
- All have same envelope

0,0   1,0

0,1   1,1

OQPSK

$-T$ 0 T

$E \cos 2\pi f_c t$

$E \sin 2\pi f_c t$

$s_{OQPSK}(t)$

ing constant envelope.[*] There are no impulsive changes at the bit termi-
nus. Phase continuity is introduced to reduce the bandwidth occupancy.
This does not imply that the spectral sidelobes go to zero, but there is a
substantial improvement because of the continuous phase across the bit
junctions even though there are still abrupt transitions (cusplike or
kinks). However, it is a constant envelope signal and does not encounter
the AM-to-PM conversion in amplifiers which would modify the spectrum if it were not constant envelope.

---

[*] Constant envelope signals are desirable so that they can work with nonlinear amplifiers, which are more efficient than linear amplifiers. This is especially important in wireless communications where conservation of battery power is important.

Figure 6.2
Power spectral densities of digital waveforms.

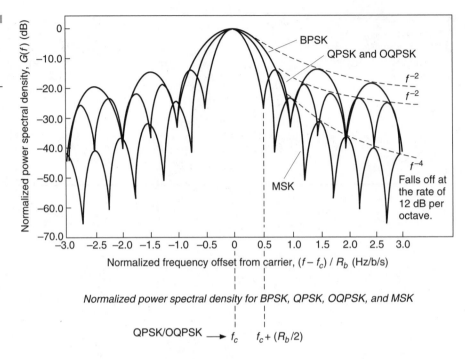

Normalized power spectral density for BPSK, QPSK, OQPSK, and MSK

$$\text{QPSK/OQPSK} \longrightarrow f_c \qquad f_c + (R_b/2)$$

6.3 MSK Modulation

One subset of CPM, which is now being considered in digital communications, is minimum shift keying (MSK). This is a binary digital FM waveform with a modulation index, $h = 0.5$. That is, Δf/bit rate $= \Delta f/R_b = \frac{1}{2}$. This can also be looked at as an OQPSK waveform with sinusoidal weighting of the data pulses as opposed to no weighting (uniformity) of OQPSK pulses. This modulation is represented in Fig. 6.3. Because the MSK is related to OQPSK, it is not surprising that the bandwidth efficiency (as defined previously) is the same, that is, 2 bits/s/Hz.

Note the weighting and the ultimate generation of the continuous phase MSK signal. The phase changes occur at the zero crossings of the weighted signal.

Digital FM modulation may take on other values of the modulation index $(\Delta f/R_b)$. We may put this in the form

$$\Delta f/R_b = n/2 \qquad\qquad [6.1]$$

where n = an integer
$\Delta f = f_1 - f_0$
R_b = bit rate $= 1/T$

Figure 6.3
Generation of the
MSK waveform.

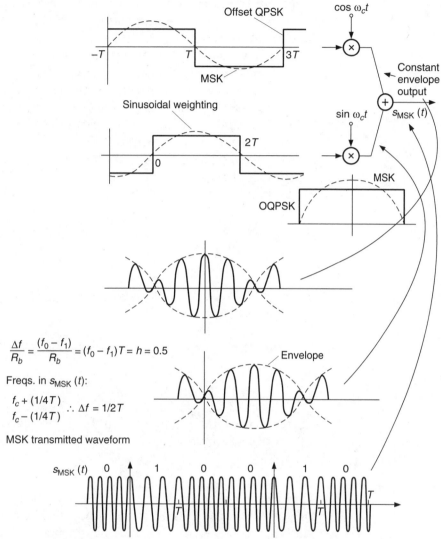

$$\frac{\Delta f}{R_b} = \frac{(f_0 - f_1)}{R_b} = (f_0 - f_1)T = h = 0.5$$

Freqs. in $s_{MSK}(t)$:

$$\begin{matrix} f_c + (1/4T) \\ f_c - (1/4T) \end{matrix} \quad \therefore \Delta f = 1/2T$$

MSK transmitted waveform

- Note continuous transitions at bit intervals. Phase continuity (piecewise) in the RF carrier at the bit transitions.
- Constant envelope signal.
- Shaping of each data bit is done over one bit interval.

For MSK, $n = 1$ or

$$\Delta f / R_b = 1/2 \qquad [6.2]$$

MSK is a special form of FSK, where the frequency shift is chosen to be exactly one-half the data rate or the frequency separation is ½*T*. Minimum shift keying alludes to the fact that the frequency separation is minimum and can still be orthogonal.[*] In BFSK the frequency separation has to be $R_b = 1/T$ for orthogonality. In Eq. [6.1], $n = 2$. Clearly, since the MSK bits are separated by one-half the frequency that BFSK is (for orthogonality), it is therefore more spectrally efficient.

The time function of an MSK signal is depicted in Fig. 6.4. Clearly this is a constant envelope signal with no impulsive phase jumps. As indicated previously, the frequency separation between the 0 and 1 bits is $f_{hi} - f_{lo} = \frac{1}{2}T$. MSK has also been referred to as fast FSK (FFSK).

The normalized power spectral density of the MSK signal is illustrated in Fig. 6.2. The main lobe is wider than that for QPSK/OQPSK, but the sidelobes fall off as $1/f^4$.

The fraction of the power located within a bandwidth B is defined as

$$P_B = \frac{\int_{-B/2}^{B/2} S(f)\, df}{\int_{-\infty}^{\infty} S(f)\, df} \qquad [6.3]$$

where $S(f)$ = power spectral density
$\quad\quad B$ = bandwidth

For example, the power spectral density for an RF BPSK pulse is given by

$$S(f) = E^2 T \left[\frac{\sin(f - f_c)\pi T}{(f - f_c)\pi T} \right]^2$$

where T is the bit duration, and f_c is the carrier. The baseband translation follows from Fourier integrals.

$$\mathfrak{F}[s(t)\exp(j2\pi f_c t)] = S(f - f_c)$$

The envelope of the spectrum of BPSK falls off as $1/f^2$. The envelope of QPSK/OQPSK also falls off at the same rate. The MSK falls off as $1/f^4$ or 12 dB per octave. As shown in Fig. 6.2, the main lobe for BPSK is broader for MSK than for QPSK/OQPSK but narrower than for BPSK. Therefore, for a given bandwidth, QPSK/OQPSK and MSK can support information at twice the rate as BPSK.

---

[*] Orthogonal signals do not interfere with each other in the process of detection. The condition for orthogonality is $\int_0^\tau s_i(t)s_j(t) = 0$ for $i \neq j$. As opposed to antipodal signals where $s_i = -s_j$. Antipodal signals (e.g., BPSK) give better error probabilities than orthogonal signals (e.g., FSK) for the same energy (or E_b/N_0). See Sec. 6.9, Signal Orthogonality, for greater details.

Figure 6.4

Minimum shift keying (MSK) modulation waveform.

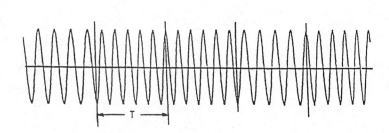

- Digital Frequency Modulation, note two frequencies.

- Note, no impulsive jump in phase (continuous).

- Constant envelope.

- The two frequencies are orthogonal (minimum separation at which this is possible), therefore $\Delta\emptyset = \pm90^0$ on signal space plane.

- $f_{hi} - f_{lo} = \Delta f = R_b/2 = 1/2T.$ *

- $h = \Delta f/R_b = \Delta fT = 0.5.$

T: symbol duration
R_b: bit rate

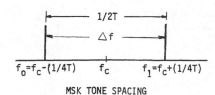

MSK TONE SPACING

* $\Delta f = f_1 - f_2 = \left[f_c + (h/2T)\right] - \left[f_c - (h/2T)\right]$

  $= (h/2T) + (h/2T)$

  if h=1/2 (MSK)

  $\therefore \Delta f = (1/4T) + (1/4T) = 1/2T$

N.B.: This should not suggest that the transmitted signal will consist of two delta functions in frequency (at $f_c \pm 1/4T$).

We can make the following observations (through calculations, that is) regarding the bandwidth at the various percentage points and the power it includes.* Frequently bandwidth takes on different bounds to different engineers. Therefore, the bandwidth one is considering should be specified. The bounds on the spectrum are indicated in the following table (see Amoroso).

---

* One should also use caution in specifying percent bandwidth since some modulations do not manifest spectral sidelobes (e.g., GMSK – $B_bT < 0.4$; TFM).

Modulation	½ Power	Null-null	99 %
BPSK	$0.88R_b$	$2.00R_b$	$20.56R_b$
QPSK/OQPSK	0.44	1.00	10.28
MSK	0.59	1.50	1.18

Bandwidth is given in terms of bit rate, R_b.

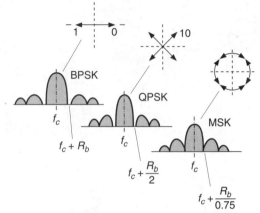

It is interesting to note that as far as spectral efficiency is concerned, the power in BPSK decays slowly and extends way out. This can interfere with a user wanting to use some contiguous spectrum, as opposed to wanting to use MSK where 99 percent of the power resides within the 1.18-bit rate R_b. Therefore, a more spectrally efficient modulation occurs. As an example, to transmit a 16-kilo bit per second data stream using the null-to-null bandwidth, the BPSK signal would require $2 \times 16 = 32$ kHz of bandwidth. Whereas, MSK would require $1.5 \times 16 = 24$ kHz bandwidth. MSK can therefore transmit 25 percent more information for the same bandwidth. Clearly, there is also an improvement in interchannel interference (ICI) and intersymbol interference (ISI). Additional information on spectral occupancy is shown in Fig. 6.5 and has been adapted from Amoroso.

6.4 Signal State Space Constellations

The signal state space diagram for MSK is shown in Fig. 6.6a. Note that the transitions occur continuously* between adjacent states. Since these are orthogonal, the phase shift is either +90° or −90° for either a 1 or 0. Also shown on the figure (Fig. 6.6) are constellations for QPSK and OQPSK. Note for QPSK (Fig. 6.6e), there are impulsive transitions from one state to another. For OQPSK (Fig. 6.6f), the transitions are similar to MSK, but the transitions are not smooth. It is also observed that neither OQPSK nor

---

* With kinks, but not discontinuities. This is a subtle point. A curve can be continuous, but not necessarily smooth. For example, /\/\ is continuous but not smooth. In MSK since the frequency within a bit is constant (f_1 or f_0), the phase is linear (and smooth) from the simple relationship $f_i = (\frac{1}{2}\pi)d\phi/dt$.

Modulation type	Half-power bandwidth	Null-to-null bandwidth	99% Energy containment bandwidth
BPSK	$0.88R_b$	$2.00R_b$	$20.56R_b$
QPSK	0.44	1.00	10.28 (90%: $0.8R_b$)
MSK	0.59	1.50	1.18 (90%: $0.8R_b$)
SFSK[1]	0.70	1.72	2.20
Quasi-Band Limited[2]	0.47	No well-defined nulls	0.95
8 PSK	0.29		6.85
16 PSK, 16 QAM	0.22		5.14
SQPSK			10.3
TFM		No well-defined nulls	0.79

BW always in units of bit rate, R_b.

(1) Sinusoidal FSK
(2) Not constant envelope

GMSK

Modulation type	35-dB bandwidth	50-dB bandwidth
BPSK	$35.12R_b$	$201.04R_b$
QPSK	17.56	100.52
MSK	3.24	8.18
SFSK	3.20	4.71
Quasi-Band Limited	1.68	2.38

B_bT	90%	99%
0.2	$.52R_b$	$0.79R_b$
0.25	.57	0.86
0.5	.69	1.04
MSK	.78	1.20
TFM	.52	0.79

MSK, TFM for comparison

Figure 6.5 Bounds on the percentage of spectral power.

Figure 6.6

Signal state space
constellation and
phase trajectories for
MSK/GMSK.

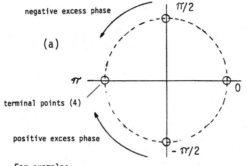

negative excess phase

$\pi/2$

(a)

π 0

terminal points (4)

positive excess phase

$-\pi/2$

For example:

MSK MODULATION SIGNAL SPACE DIAGRAM

- The four data points at $\emptyset=0, \pi/2, \pi$ & $3\pi/2$ represent the terminal point phase states.
- Sine wave shaping causes <u>gradual linear $\pm90°$ phase transitions.</u>
- Inherently maintains phase continuity from bit to bit.
- Because the two states are orthogonal, the shift in frequency translates to either a $+90°$ phase shift or $-90°$ phase (for either a 1 or 0 data input to the modulator).
- Actually CPM signals cannot be represented by discrete points in signal space like for PSK because the carrier phase is dynamic. Therefore we revert to phase trajectories to illustrate the signal phase as a function of time.

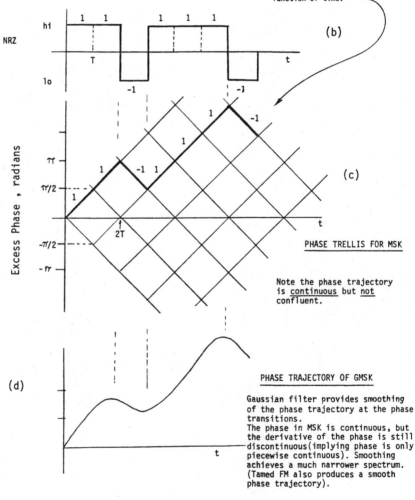

(b)

NRZ

hi   1   1   1   1   1

T   t

lo   -1   -1

(c)

Excess Phase , radians

π

$\pi/2$

2T

$-\pi/2$

$-\pi$

PHASE TRELLIS FOR MSK

Note the phase trajectory is <u>continuous</u> but <u>not</u> confluent.

(d)

t

PHASE TRAJECTORY OF GMSK

Gaussian filter provides smoothing of the phase trajectory at the phase transitions.
The phase in MSK is continuous, but the derivative of the phase is still discontinuous(implying phase is only piecewise continuous). Smoothing achieves a much narrower spectrum. (Tamed FM also produces a smooth phase trajectory).

Figure 6.6
(continued)

(e)

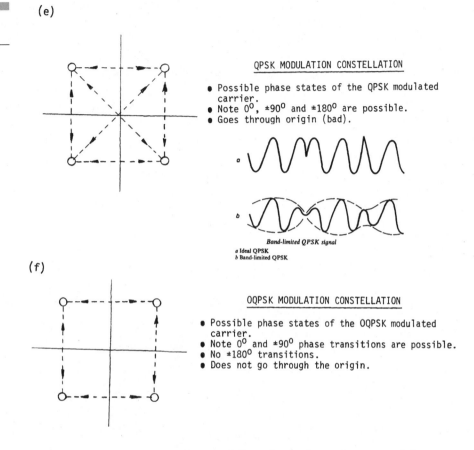

QPSK MODULATION CONSTELLATION

- Possible phase states of the QPSK modulated carrier.
- Note 0^0, $\pm90^0$ and $\pm180^0$ are possible.
- Goes through origin (bad).

Band-limited QPSK signal
a Ideal QPSK
b Band-limited QPSK

(f)

OQPSK MODULATION CONSTELLATION

- Possible phase states of the OQPSK modulated carrier.
- Note 0^0 and $\pm90^0$ phase transitions are possible.
- No $\pm180^0$ transitions.
- Does not go through the origin.

MSK go through zero. QPSK transitions do go through zero and thus produce 180° jumps in phase, although obviously not a desirable feature.

6.5 GMSK and TFM Modulations

It was noted that MSK has several desirable attributes, but there is still out-of-band spectrum which can be deleterious to performance. The MSK spectrum can be controlled by the use of a premodulation (FM) low-pass filter as shown in Fig. 6.7,* having a gaussian passband character-

---

* It is significant to note that the sketch at the top of Fig. 6.7 indicates that a voltage-controlled oscillator is keyed as opposed to keying in two oscillators (to produce f_{HI} and f_{LO}). The former produces a continuous phase modulation, where the latter would produce a discontinuous phase at the bit transitions.

istic. Figure 6.7 shows the power spectra of GMSK where the normalized
3-dB bandwidth (B_b) of the premodulation gaussian LPF ($B_b T$) is parametric. It is riveting to observe the increase in spectral efficiency which
results for various values of $B_b T$. For small values of $B_b T$, the sidelobes are
actually insignificant. The value of $B_b T = 0.3$ is particularly interesting
because it is a spectrum which has been produced by a different processing partial response signaling of the data bits into the FM modulator, that

Figure 6.7

Power spectra of
GMSK for parametric
values of the normal-
ized bandwidth $B_b T$.

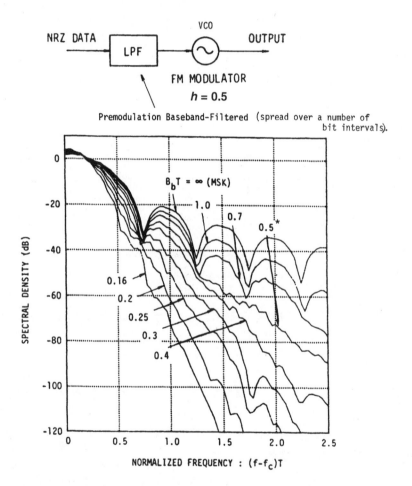

- B_b: Bandwidth of Gaussian filter.
- $B_b T$=0.3 is value used in GSM (European).
- $B_b T$=0.2 gives the approximate spectrum for Tamed FM.
- * $B_b T$=0.5 indicates that spreading is over two bit intervals.

This bit spreading reduces
the spectral spread and
thus the ACI.

is, a scheme called tamed FM (TFM). Continuing in this vein, 99 percent of the spectral power occupied by TFM occurs within $0.79R_b$. For MSK this value is $1.2R_b$ and a marked improvement.

These curves reflect constant envelope signals. In band limiting, which is frequently used, one can realize better spectral falloff, but this would produce nonconstant envelope signals. Here channel linearity becomes important. In fact, if linearity prevails in the channel, better performance may be possible with the use of QPSK and filtering to band limit. The linear channel will not regenerate the sidelobes after filtering. For example, spectrum containment for a band-limited signal is shown in Fig. 6.5 and can restrict the out-of-band emissions.

Figure 6.8 illustrates the measured spectra of the RF signal output for parametric values of B_bT. The test results agree fairly well with the plots shown in Fig. 6.7. There are insignificant spectral sidelobes for small values of B_bT. However, there is some degradation in the bit error rate for the same E_b/N_0 compared to BPSK. This is understandable since there is a greater truncation of the spectrum (and thus the power in the signal). From Fig. 6.7, observe that some energy is also taken out of the main lobe. This reduces the power efficiency.

NOTE: *The European cellular system uses GSM with a value of* $B_bT = 0.3$ *Hz/b/s, where B is the bandwidth of the gaussian filter. It is safe to say that GMSK will be used extensively in future wireless systems where spectral efficiency is of paramount importance.*

6.6 Continuous Phase Modulation

MSK is a special case of a large class of constant amplitude modulation schemes called continuous phase modulation (CPM). As shown in previous sections, the minimum shift keying (MSK) can be considered a special case of FSK (CPM) with a frequency separation $(f_1 - f_2)$ of $\frac{1}{2}T$, where T is the bit interval.

$$\Delta fT = (f_1 - f_2)T = 1/2$$
$$\Delta fT = 1/2$$
$$\Delta f/R_b = 1/2$$

where R_b = bit rate

Figure 6.8
Measured power
spectra of GMSK for
parametric values of
B_bT.

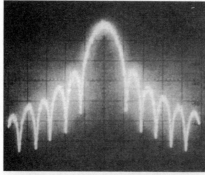

$B_bT = \infty$
(MSK)

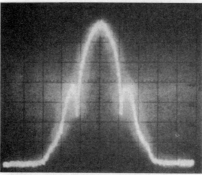

$B_bT = 0.5$

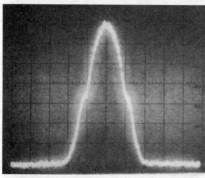

$B_bT = 0.25$

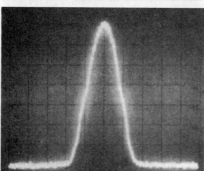

$B_bT = 0.2$

A baseband digital signal modulated onto a carrier can be given by the general expression for an FM waveform. This relationship stems from FM theory long before digital was even considered (see van der Pol).

The frequency modulated (a CPM signal) carrier can be represented by

$$v_{FM}(t) = E \cos [\omega_c t + \phi(t)] \qquad [6.6]$$

where the phase function $\phi(t)$ is related to the information or baseband signal waveform (analog or digital)

$$\phi(t) = 2\pi h \int_{-\infty}^{t} v_m(\tau) \, d\tau \qquad [6.7]$$

where h = modulation index
v_m = frequency modulating signal

$$v_{FM}(t) = E \cos \left[\omega_c t + 2\pi h \int_{-\infty}^{t} v_m(\tau) \, d\tau \right]^{\textbf{.}} \qquad [6.8]$$

The expression in brackets represents the total instantaneous phase of the modulated signal. The instantaneous frequency is

$$f_i \triangleq (1/2\pi) d\phi/dt = f_c + hv(t) \qquad [6.9]$$

Unless $v(t)$ contains an impulse, the FM signal $v_{FM}(t)$ is continuous phase (CPFM).

In the digital world, we may now consider the input to be a series of pulses and the output CPM signal takes the form (see Gronemeyer and McBride)

$$v_{CPM}(t) = E \cos \left[2\pi f t + 2\pi h \int_{0}^{t} \sum_{m=0}^{\infty} a_n g(\tau - nT) \, d\tau \right] \qquad [6.11]$$

where h = modulation index
$a_n = \pm 1, \pm 3, \ldots \pm (M - 1)$, M possible symbols (we will consider binary only, i.e., ± 1)
$g(t)$ = frequency pulse with finite duration

and where the *information*-carrying phase $\phi(t)$ is the integral term of Eq. 6.11.

$^{\textbf{.}}$ From classical FM theory, it is recalled that if v_m is a simple sinusoid $v_m = E \cos \omega_m t$, the FM wave becomes $v_{FM} = E \cos (\omega_c t + \beta \sin \omega_m t)$, where β is the familiar modulation index and $\beta = \Delta\omega / f_m = \Delta f / f_m$.

In GMSK $q(t)$ is chosen to be a gaussian pulse, whose normalized bandwidth is

$$B_b T = x \qquad (\text{e.g., GSM, } B_b T = 0.3)$$

The phase term given in 6.11 may also be expressed in terms of the baseband phase pulse $q(t)$ defined by:

$$q(t) = \int_{-\infty}^{t} g(\tau)\, d\tau$$

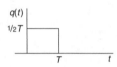

$g(t)$ is normalized such that

$$\int_{-\infty}^{\infty} q(t)\, dt = \tfrac{1}{2} \text{ (area under pulse equal to 1/2)}$$

or

$$q(\infty) = \tfrac{1}{2} \qquad\qquad\qquad [6.12]$$

Therefore Eq. 6.11 becomes

$$v_{\text{CPM}} = E\cos\left[2\pi f + 2\pi h \sum_{n=0}^{\infty} a_n q(t - nT)\right] \qquad [6.13]$$

The transmitted M-ary symbols a_n appear in the phase

$$\phi(t) = 2\pi h \sum_{n=0}^{\infty} a_n q(t - nT) \qquad\qquad [6.14]$$

where $q(t)$ is the shaping pulse, and ramping of $q(t)$ from 0 to T yields a linear phase transition of $\pi/2$.

As an example, consider $g(t)$ as a display of rectangular binary pulses as shown in Fig. 6.9b. The pulse widths are T and $h = \tfrac{1}{2}$. We normalize $g(t)$ such that the area under each pulse is $\tfrac{1}{2}$. Therefore, for each pulse we have

$$(\tfrac{1}{2})(2\pi h) = \pi h = \pi(\tfrac{1}{2}) = \pi/2 \qquad\qquad [6.15]$$

Therefore, each pulse yields a 90° phase change at each data integral, a requirement for orthogonality. We have exercised MSK in this section.

6.7 Tamed Frequency Modulation

The generalized circuit for digital FM is shown in Fig. 6.9a. It is the premodulation filter $g(t)$ which determines the spectral efficiency. The spectrum clearly must be as narrow as possible. Notice that it is a function of the value of $h \cdot (f_1 - f_2) T$.

Figure 6.9
Continuous phase
modulation (digital
FM).

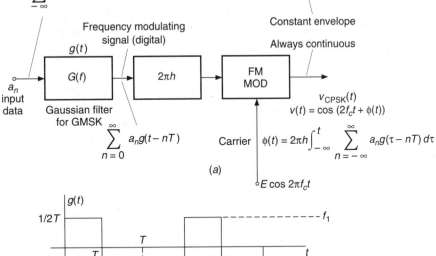

Data stream:

$$a(t) = \sum_{-\infty}^{\infty} a_n \delta(t - nT) \quad a_n: \pm 1$$

Therefore can be amplified by efficient
nonlinear power amplifiers. Desirable
in wireless communications where
battery power is precious.

Frequency modulating
signal (digital)

$g(t)$

Constant envelope

Always continuous

a_n input data — $G(f)$ — $2\pi h$ — FM MOD — $v_{CPSK}(t)$

$v(t) = \cos(2f_c t + \phi(t))$

Gaussian filter for GMSK

$$\sum_{n=0}^{\infty} a_n g(t - nT)$$

Carrier $\quad \phi(t) = 2\pi h \int_{-\infty}^{t} \sum_{n=-\infty}^{\infty} a_n g(\tau - nT) \, d\tau$

$E \cos 2\pi f_c t$

(a)

$g(t)$

$1/2T$ .... f_1

T

T

t

.... f_2

• Rectangular pulses of width T.
• Each pulse has an area equal to 1/2.

$f_1 = f_c + (h/2T)$

$f_2 = f_c - (h/2T)$

$\Delta f = f_1 - f_2 = f_c + (h/2T) - (f_c - (h/2T))$

$\quad = (h/2T) + (h/2T)$

(b)

if $h = 1/2$ (MSK):

$$\Delta f = [f_c + (1/4T)] - [f_c - (1/4T)] \quad \therefore \quad \boxed{\Delta f = (1/4T) + (1/4T) = 1/2T}$$

Note, this is the minimum frequency difference
that will provide orthogonal signals. That is,
the frequency difference between the "1" bit
and "0" bit is 1/2 the bit rate, $R_b/2$.

It was observed that even though MSK is a continuous phase modulation, there are still undesirable spectral sidelobes.[*] It is true in signal state space that the phase changes for MSK are more gradual than those mani-

---

[*] The main advantage of MSK is its spectral efficiency. Ninety-nine percent of the power will occupy a bandwidth equal to 1.2 times the bit rate R_b.

fested by either BPSK or QPSK/OQPSK; it does not completely eliminate the sidelobes. There are other variations of CPM which can improve this situation still further. One, in particular, which is elegant and interesting, has been labeled tamed frequency modulation (TFM) (see de Jager and Dekker). In the cited paper, de Jager and Dekker spread the 90° phase change over three bit intervals (partial response signaling). For example, if the data is defined by

$$a(t) = \sum_{n=-\infty}^{\infty} a_n(t - nT) \qquad a_n = +1 \text{ or } -1 \qquad [6.16]$$

If we were considering MSK, we expect the phase change during one bit interval to be 90°. Therefore, for MSK

$$\phi(mT + T) - \phi(mT) = (\pi/2)a_m \qquad [6.17]$$

For TFM the rule is

$$\phi(mT + T) - \phi(mT) = (\pi/2)[a_{m-1}/4 + a_m/2 + a_{m-1}/4]$$

with $\phi(0) = 0$ if $a_0a_1 = +1$, $\qquad \phi(0) = \pi/4$ if $a_0a_1 = -1$ $\qquad [6.18]$

Note that the phase is spread over three bit intervals.

If the three sequential bits have the same polarity, the phase change is 90°, and the phase remains constant if the three succeeding bits have alternating polarities.

In summary, on the discussion of TFM, we have exercised Eq. (11). Repeating Eq. (11) for CPM and for binary waveforms, we are considering here $(a = \pm 1)$. We have

$$v_{CPM} = E \cos\left[2\pi ft + 2\pi h \int_0^t \sum_{n=0}^{\infty} a_n g(\tau - nT) \, d\tau \right] \qquad [6.11]$$

$$\int_{-\infty}^{\infty} g(t) \, dt = \frac{1}{2} \qquad a_n = \pm 1$$

In general, if $h = \frac{1}{2}$, the total contribution to the phase integral $\phi(t)$ for each pulse is $g(t) = 90°$. For *TFM* this 90° is spread out over several symbols, if $g(t)$ lasts for more than one symbol. For $h = \frac{1}{2}$ (and binary case) you have to end up with the integral of $g(t) = \frac{1}{2}$.

The premodulation transfer characteristic of the filter is more complicated than that used for MSK (sinusoidal weighting) or GMSK (gaussian weighting). In one implementation, the baseband modulating pulse $g(t)$ consists of a transversal filter followed by a Nyquist III filter (see Nyquist).

The mathematics are somewhat complicated, but the reader may refer to the article for further information.

The spectrum for the TFM signal is shown in Fig. 6.7. The curve labeled $B_b T = 0.2$ is the approximate spectrum for TFM. Note that it has no sidelobes like certain values of GMSK. However, the spectrum is more noticeable than in MSK. There is a small loss in communication efficiency, but the lack of sidelobe emissions is worth the price. The shortfall in error performance (about 1 dB) can be made up in channel coding (e.g., FEC).

6.8 π/4-QPSK Modulation

Another modulation which is receiving wide acceptance in wireless systems for both satellite and mobile wireless applications is π/4-QPSK. Its constellation in the signal state phase plane is depicted in Fig. 6.10. Note that the carrier phase transitions are restricted to ±45° and ±135°. The modulation uses eight phases to transmit two bits of information. The two bits are mapped into one of four phase transitions from one symbol to the next. The ±π transitions found in QPSK do not exist. This can be observed by referring to the signal state space diagram illustrated in Fig. 6.11a. The transitions can occur from any state to any other.

π/4-QPSK uses Nyquist baseband signals (35 to 50 percent rolloff), and they are not constant envelope. Generation would normally require linear amplifiers, but nonlinear efficient power amplifiers can be used and require negative feedback control to reduce distortion. Since the carrier phase does not undergo ±180° transitions, the envelope fluctuations after filtering are not excessive and subsequent subjection to nonlinearities (AM to PM) will not significantly restore the sidelobes. Any restoration that may exist can be further mitigated by using negative feedback as indicated previously.

The dynamic display of the signal state space constellation for QPSK and π/4-QPSK are shown in Fig. 6.11. The QPSK signals, in going from one state to another, can go through zero or ±180° phase transitions. In Fig. 6.11b (π/4-QPSK) the transitions avoid going through zero. Figure 6.11c shows the π/4-QPSK constellation displayed on an ANRITSU/WILTRON digital mobile radio transmitter tester. As indicated previously, there are only four phase transitions from each symbol (±45°, ±135°).

π/4-QPSK's attractive features have lent itself to be used in the U.S. cellular radio standard IS-54 and in the Japanese cellular system, personal digital cellular (PDC).

Figure 6.10
The π/4 constellation
on signal space
diagram.

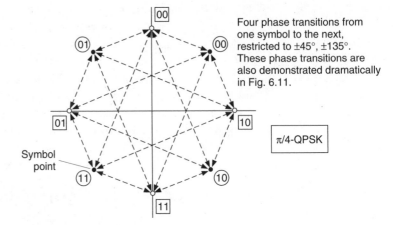

Four phase transitions from
one symbol to the next,
restricted to ±45°, ±135°.
These phase transitions are
also demonstrated dramatically
in Fig. 6.11.

π/4-QPSK

• Two bits are transmitted for each symbol represented by a transition to a new state.

• Does not go through zero in phase plane, for example, no ±180° transitions.

• Possible phase shifts: ±45°, ±135°, link between states indicates the allowed transitions.

• Not constant envelope modulation.

• Less degradation due to nonlinearity than for QPSK. Same spectrum as OPSK.

• With negative feedback to reduce distortion, permits use of nonlinear power amplifiers to increase efficiency (see Akaiwa and Nagata).

• Used in IS-54 (USA), replace 30 kHz (AMPS) with three channels.

• Eight phases are possible, but only four are active at any one time.

• Appears to be an 8-phase pattern, but it is not. Signal elements of the modulated signal are selected in turn from two QPSK constellations which are shifted by π/4 with respect to each other.

Like in previous modulations discussed, the spectral efficiency results from reducing the spectral sidelobes and thus reducing ISI and ACI.

6.9 Signal Orthogonality

Signal orthogonality is achieved if the correlation coefficient between the two signals equals zero.

$$\rho = \int_0^T s_1(t)s_0(t)\ dt = 0 \qquad \text{[6.18']}$$

If we assume the signaling states, normalized to unit energy, are represented by

Figure 6.11
Measured signal state space constellations.

$$s_1(t) = \sqrt{2/T} \cos 2\pi f_1 t \qquad \text{[6.19]}$$

$$s_2(t) = \sqrt{2/T} \cos 2\pi f_0 t \qquad 0 \leq t \leq T \qquad \text{[6.20]}$$

Equation [6.18'] becomes

$$\rho = \int_0^T (\sqrt{2/T} \cos 2\pi f_1 t \times \sqrt{2/T} \cos 2\pi f_0 t) \, dt \qquad \text{[6.21]}$$

Using the trigonometric identity

$$\cos \alpha \cos \beta = (\tfrac{1}{2}) \cos (\alpha - \beta) + (\tfrac{1}{2}) \cos (\alpha + \beta) \qquad \text{[6.22]}$$

plugging in Eq. [6.21], we obtain

$$\rho = \int_0^T (1/T) [\cos (2\pi f_1 - 2\pi f_0)t + \cos(2\pi f_1 + 2\pi f_0)t] \, dt \qquad \text{[6.23]}$$

integrating and plugging in the limits, we obtain:

$$\rho = (1/T)[\tfrac{1}{2}\pi(f_1 - f_0)] \sin [2\pi(f_1 - f_0)]t \big|_0^T$$

$$+ (1/T)[\tfrac{1}{2}\pi(f_1 + f_0)] \sin [2\pi(f_1 + f_0)]t \big|_0^T$$

$$= \frac{\sin [2\pi(f_1 - f_0)T]}{2\pi(f_1 - f_0)T} + \frac{\sin [2\pi(f_1 + f_0)T]}{2\pi(f_1 + f_0)T} \qquad \text{[6.24]}$$

If $(f_1 + f_2) \gg (f_1 - f_2)$, Eq. [6.24] reduces to

$$\rho = \frac{\sin [2\pi(f_1 - f_0)T]}{2\pi(f_1 - f_0)T} = \frac{\sin [2\pi\Delta f T]}{2\pi\Delta f T} \qquad \text{[6.25]}$$

where $\Delta f = f_1 - f_0$ is the separation between high and low frequencies.

Equation [6.25] is shown plotted in Fig. 6.12, which is recognized of the form $\sin x/x$.

We notice that when

$$\Delta f = 1/2T, 1/T, 3/2T, \ldots, n/2T$$

the function goes to zero, and the correlation equals zero. The case where $\Delta f T = \Delta f/R_b = \tfrac{1}{2}$ is for minimum shift keying (MSK), and this is the closest frequency separation for orthogonality.

It can be shown that the optimum probability of error of FSK occurs when the correlation function is negative. In this display, when $\rho = -0.212$, this results from the probability of error relationship for equal energy signals.

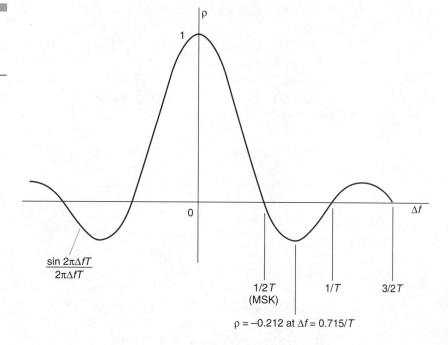

Figure 6.12
Correlation showing regions of signal orthogonality.

$$P(e) = Q(\sqrt{(E/N_0)(1-\rho)})$$
$$= Q(\sqrt{1.22(E/N_0)})$$

Therefore, the optimum FSK is +0.86 dB better than orthogonal FSK. Note that $P(e)$ for an antipodal signal (e.g., BPSK) becomes

$$P(e)Q(\sqrt{2(E/N_0)})$$

since in this case $\rho = -1$, which gives a $P(e)$ which is 3 dB better than MSK. For orthogonal signals ($\rho = 0$), the performance is worse than antipodal signals by 3 dB.

6.10 Summary and Conclusions

In this chapter, we have addressed the concept of spectral efficiency in modern digital communications. Spectral efficiency is defined as the minimum spectral content of a signal which gives adequate bit error rates (BER), and, in addition, signal formats which reduce the out-of-band inter-

ference allowing closer channel packing and reduced intersymbol interference.

The classical formats (PSK and QPSK) are still popular and will continue to be used. But on the downside, they provide out-of-band interference because of their spectral sidelobes (see Fig. 6.2). In practice, these are
suppressed by filtering, but filtering takes away their constant envelope
attribute. In fact, the envelope goes to zero when there are 180° phase transitions between the data bits. This therefore requires the use of linear
amplifiers, in lieu of the more efficient nonlinear power amplifiers. In
wireless communications, this places a burden on the power supply, and
even more so for handheld transceivers.

Some of the advanced modulation techniques discussed in this chapter
are able to alleviate some of the previously cited problems by using modulation schemes such as MSK, GMSK, and other continuous phase modulations. Their benefit stems in part from the fact that the phase
transitions between symbols are not impulsive but continuous. One step
in the right direction is the use of offset QPSK (OQPSK), where phase
changes do not exceed 90°. These reduced phase transitions in CPM have
the desirable effect of reducing the spectral sidelobes still further. These
are constant envelope signals, and any additional filtering will essentially
maintain their constant envelope. Nonlinear amplifiers may be used.

It is this author's opinion that these new modulation techniques will
find increased use in wireless and satellite communications applications
because of the paucity of spectrum. To single out one particular modulation, GMSK will find the most increased use.

References

Akaiwa, Y., and Y. Nagata, "Highly Efficient Digital Mobile Communications with a Linear Modulation Method," *IEEE Journal on Selected Areas
in Communications,* June 1987.

Amoroso, F., "The Bandwidth of Digital Data Signals," *IEEE Com. Magazine,*
November 1980.

Bennett, W. R., and J. R. Davey, *Data Transmission,* McGraw-Hill Book Co.,
New York, 1964.

Cuccia, C. L., *Harmonics, Sidebands and Transients in Communication Engineering,* McGraw-Hill Book Co., New York, 1952.

de Jager, F., and C. B. Dekker, "Tamed Frequency Modulation—A Novel Method to Achieve Spectrum Economy in Digital Transmissions," *IEEE Com. Magazine,* 1978.

Gronemeyer, S. A., and A. L. McBride, "MSK and Offset QPSK Modulation," *IEEE Trans. on Communications,* August 1976.

Kalet, I., Course notes from a course given at George Washington University on Modern Digital Communications, Summer 1994.

Murita, K., et al., "GMSK Modulation or Digital Mobile Radio Telephony," *IEEE Trans. on Communications,* July 1991.

Nyquist, H., "Certain Topics in Telegraph Transmission Theory," *Trans. AIEE,* vol. 47, February 1928.

Pasupathy, S., "Minimum Shift Keying: A Spectrally Efficient Modulation," *IEEE Com. Magazine,* July 1979.

Sue, M. K., and Y. H. Park, "Second Generation Mobile Satellite System," JPL Pub. 85-58, June 1985.

Sundberg, C. C., "Continuous Phase Modulation," *IEEE Com. Magazine,* April 1986.

van der Pol, B., "Frequency Modulation," *Proc. IRE,* July 1930.

Antenna Subsystems for Mobile Satellite Communications

7.1 Introduction

Earth terminals which are viewing low earth satellites can see them over an extremely wide swath even if only for relatively short periods of time. A representative scenario of the earth terminal location and an orbiting vehicle is depicted in Fig. 7.1. To simplify the geometry, the satellite is shown overlaying the terminal or through its zenith. The earth radius is given as the equatorial radius of 6378 km. The one-half central angle is given as θ and altitude h. The central angle, which is one-half the viewer's central angle, is given as

$$\theta = \arccos\left[(R_E/R_E + h)\cos\alpha\right] - \alpha \qquad [7.1]$$

where α = elevation angle of earth terminal antenna axis to the satellite
R_E = earth radius

Clearly, when $\alpha = 90°$, the satellite is at the zenith position, and $\theta = 0°$.

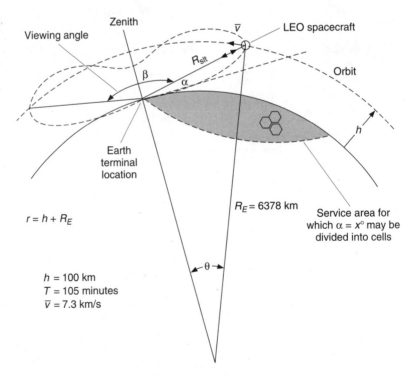

Figure 7.1
A LEO spacecraft-earth station interface scenario.

The slant range to the satellite is R_{slt} and from the law of cosines we obtain:

$$R_{slt} = (R_E^2 + (R_E + h)^2 - 2R_E(R_E + h) \cos \theta)^{0.5} \qquad [7.2]$$

From Fig. 7.1 note that the maximum viewing angle in which the terminal can see the satellite is when the elevation angle α is equal to 0. However, this may not be the minimum optimum angle. In practice, 5 to 10° is more practical because the multipath effects and antenna radiation patterns at low-elevation angles may not be adequate.

The acceptable elevation angle also determines the satellite's field of view which in turn dictates the maximum bounds of the cellular areas served by the satellite. For an altitude of 1000 km, and $\alpha = 10°$, the service area has a radius of about 2350 km. For the same elevation angle α and an increase in altitude, the area clearly gets larger.

The elevation angle is given by

$$\alpha = \arccos [(R_E + h) \sin \theta / R_{slt}] \qquad [7.3]$$

or

$$\cos \alpha = [(R_E + h)/R_{slt}] \sin \theta \qquad [7.4]$$

since $\cos \alpha$ has to be equal to 1 for $\alpha = 0°$, $\sin \theta$ must be equal to $R_{slt}/(R_E + h)$, and the earth station viewing angle is therefore 180°. For the practical elevation angle of 10°, the viewing angle β is 160°. This is a rather wide viewing angle for an antenna to straddle, considering that the antenna beam width must be that wide in all directions since generally the satellite will not fly over the earth station. Ideally, the earth station antenna pattern must be nearly hemispherical with some sacrifice in gain permitted in the zenith area, or the minimum slant range must be acceptable since this is where the satellite is closest to the earth terminal. It is possible to shape the antenna pattern such that the satellite's power-flux density (W/m^2) at the earth station is nearly constant as the satellite comes closer to the earth station.

For an elevation angle of 10° and a satellite altitude of 1000 km, the central angle θ from Eq. [7.1] is equal to 2°.

From Eq. [7.5], at $\theta = 23°$, the slant range of the satellite from the earth terminal is:

$$R_{slt} = 2920 \text{ km}$$

For the satellite at zenith, the slant range $(R_{slt10}°)$ is 1000 km. The slant range loss ratio for the satellite at the two ranges is therefore:

$$L = 10 \log (R_{\text{slt10}}°/R_{\text{slt10}}°)^2$$
$$= 10 \log (2920/1000)^2$$
$$= 9 \text{ dB} \qquad\qquad\qquad [7.5]$$

This is markedly different from a satellite at GSO where the differential loss between the nadir distance and 10° elevation angle is 1.1 dB.

7.2 Earth Terminal Antenna Requirements

As indicated previously, the slant range between the satellite and the earth station is much less when it is directly above the station than when the satellite is near the horizon. In order to maintain continuous contact with the satellite, the earth station antenna must have a very broad beam width, which essentially covers all of orbital space. If satellites are observed in several orbits, the antenna pattern must be omnidirectional and hemispherical in elevation to account for all azimuth and elevation positions.

Since there is a nominal differential 9-dB signal level reaching the earth station from the satellite (at $h = 1000$ km), the signal from the satellite is maximum when near zenith, or the satellite is perpendicular to the orbit in the earth station direction. The variation in path loss as a function of the elevation angle is shown in Fig. 7.2. Therefore, in order to receive a constant level signal at the earth station, the satellite antenna pattern should have the same concave shape as the loss curve, as shown in Fig. 7.3.

The transmit satellite antenna beam therefore transmits uniform power flux density (watts per meter square) onto the earth's surface. This is frequently referred to as isoflux power density. The energy received at the earth station is independent of where the earth station is within the beam footprint. This is because the beam is shaped to produce a concave pattern which compensates for the additional losses that occur as the slant range increases as the receiver is located away from the nadir point or a point of minimum range. In other words, the satellite downlink EIRP is constant across its beam.

Actually, there may be a similar shaped beam for the satellite receive antenna. In this case, the figure of merit (antenna gain over system temperature G/T_s) changes across the beam.

In order to maintain continuous communications with the satellite, the ground receiver antenna beam width must be broad, and antenna

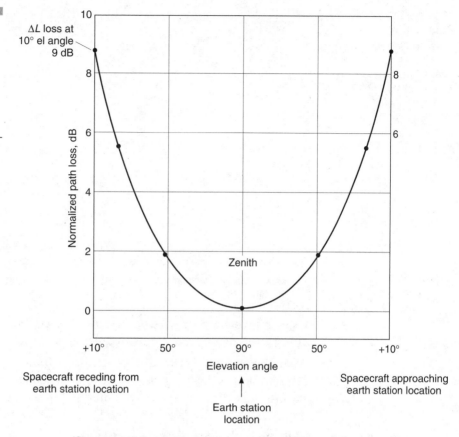

Figure 7.2
Differential path loss which occurs as the spacecraft passes from horizon to horizon (10° elevation angle) over the earth station.

ΔL loss at 10° el angle 9 dB

Normalized path loss, dB

Zenith

+10°   50°   90°   50°   +10°

Elevation angle

Spacecraft receding from earth station location

Spacecraft approaching earth station location

Earth station location

(Shape of radiation pattern to produce constant field strength on the earth)

- Normalized path loss is the loss which occurs over and above the free space loss which occurs when the spacecraft is at the earth station's zenith (at which point the minimum space loss occurs).

- This space loss can be obviated by the earth station if its antenna pattern approximates this contour.

theory indicates that the gain will be small. In addition, to make the antenna insensitive to ionospheric polarization rotation, the beam should be circularly polarized. There are several accrued reasons for using circular polarization. They are as follows.

■ No orientation of the receive or transmit antenna is required, save pointing the antenna at the satellite or ground receiver.

■ Its polarization is insensitive to Faraday polarization rotation as it negotiates the ionosphere. As is well known, linear polarization wit-

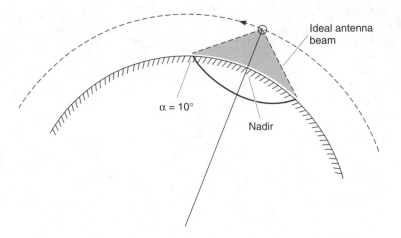

Figure 7.3
Isoflux power density satellite antenna beam. Both transmit beam (constant EIRP) and receive beam (constant G/T_s) may assume this shape.

- Uniform coverage along the surface of the earth compensating for range losses.
- The pattern is rotationally symmetrical with minimum gain at the nadir location.

nesses this rotation, can be significant for frequencies below 10 GHz, and is inversely dependent on frequency.

■ Circular polarization can mitigate multipath reflections when subjected to odd bounce off surrounding objects. The sense of polarization is reversed, and the receiver antenna is less insensitive to the wrong sense circular polarization. The carrier-to-multipath ratio improves with increasing elevation angle.

To realize circular polarization, the antenna must assume the following characteristics which are depicted in Fig. 7.4.

■ Two electric vectors must be generated which must be spatially orthogonal to each other.

■ The vectors must be equal in magnitude. If they are not equal, elliptical polarization will be produced.

■ The vectors must be in time quadrature.

There are basically two types of antennas used in mobile communications which are different in performance. There are those which have omnidirectional azimuth coverage with hemispherical elevation coverage (cardioidal-type pattern). These are axial beams. The elevation pattern may have a notch on-axis for certain antennas and have been referred to as conical patterns. For a zenith-pointing antenna, this is acceptable since the higher gain is required at lower elevation angles as described previ-

Figure 7.4

Circular polarization generation.

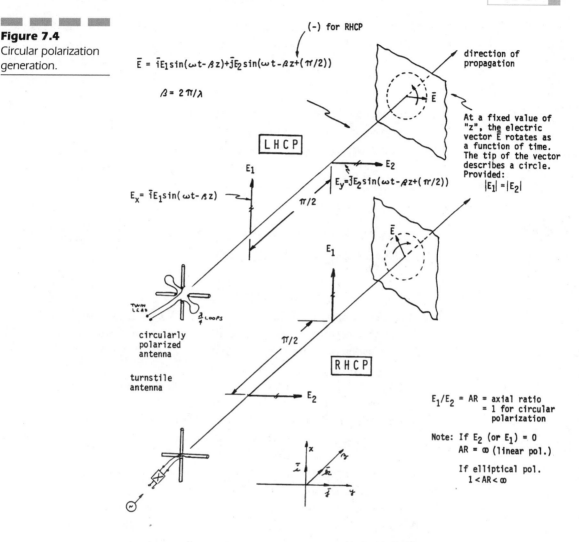

ously. These are low-gain antennas in the range of 0 to 5 dB. As indicated previously, its merit is that it can pick up satellite signals from any direction.

The other category of antennas have medium-gain (10 to 15 dB) directional beams. Several types of wire antennas, such as Yagi-Uda, log-periodic, and helices, can provide this gain, but they are high-profile structures and will find use in the larger mobiles. Therefore, we will confine our attention to low-profile array structures used on vehicle and aircraft mobiles. These have been researched extensively in recent years for mobile terrestrial and airborne use.

Directionality suggests that some kind of steering of the beam is necessary to continuously searchlight the satellite. This steering may be performed either by mechanical or electronic means. Mechanically, the antenna is slewed until a signal is received and then made to track and lock on the satellite by additional sensors. These may be closed-loop sensors (such as monopulse which use the received signal) or open-loop sensors (such as magnetic compasses or gyros). In electronic beam steering, the beam is moved by phase shifters in the antenna element paths. See Secs. 7.3.7 and 7.3.8 for further discussion of array structures.

7.3 Antenna Types Used in Mobile Satellite Service

Antenna types frequently used in mobile satellite service for LEO and GSO applications are indicated in Fig. 7.5. All of these antennas can be designed to accept or transmit circular polarization, save the whip which is linearly polarized. Its use would incur a 3-dB loss in signal since the downlink will be circularly polarized.

The types being considered for satellite-based cellular systems are the quadrifilars and broadband microstrip patches.

The mobile terminals (vehicles) would include the classical Kraus helix, short backfire dish feed, passive arrays, and phased arrays. This group, of course, would require beam steering.

The *fixed* terminal can yield high gains with an elaborate mechanical steering, if it is to follow the LEO orbit. For GSO application, beam steering may not be necessary if the inclination remains close to 0°.

A listing of some of these antennas follows. These antennas are frequently used in small earth terminals and handheld transceivers.

- Handheld terminals (low gain)
  - Whip (loaded)
  - Bifilar and quadrifilar helices
  - Broadband microstrip patches
  - Cavity-backed spirals
- Mobile terminals (medium gain)
  - Classical helix (Kraus)
  - Mechanically steered array (small)

Figure 7.5
The concept of backfire and endfire in antennas.

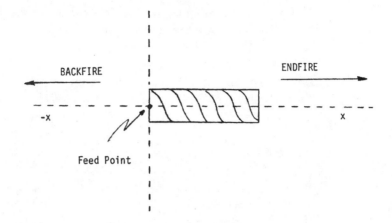

- The unifilar helix (Kraus) is fed against a ground plane.

- The bifilar and quadrifilar may work with or without a ground plane depending on whether it is fed from the bottom or top of helix respectively.

  - ▪ Electronic phased array
  - ▪ Short backfire dish
- ▪ Fixed terminals (high gain)
  - ▪ Yagi-Uda, single or array
  - ▪ Helix, single or array
  - ▪ Mechanical steered array
  - ▪ Aperture antennas
- ▪ Dipoles (and others listed previously) may be used in suitcase-size terminals.

7.3.1 Bifilar and Quadrifilar Helix Antennas

Bifilar and quadrifilar helical antennas have received considerable attention in recent years. For example, they are used in ground receivers for GPS navigational systems, for spacecraft, and on ground earth terminals (handheld) in LEO telecommunications satellite systems. They have also been used on amateur satellites.

The antenna differs from the classical Kraus helix in several ways. The Kraus helix is normally a single multiturn helix wound on a dielectric core, or if the coils are rigid enough, on an air core.

The bifilar antenna has two windings on a single core, where the quadrifilar antenna has four windings equally spaced around a core. Both the bifilar and quadrifilar antennas (henceforth referred to as bifilars and quadrifilars) may be partial turns or multiple turns. The Kraus helix works against a ground plane where the filar antennas can work without a ground plane. This feature is desirable in certain applications.

The Kraus helix intrinsically produces circular polarization, and the sense depends on the winding direction. That is, a winding wound as a right-hand screw will produce right-hand circular polarization, whereas the filar antenna wound clockwise will produce counterclockwise or left-hand circular polarization.

The Kraus helix produces an antenna pattern which is endfire. For bifilars or quadrifilars, the radiation can be either endfire or backfire, depending on how the coil elements are fed. The concept of rotation is illustrated in Fig. 7.5.

The quadrifilar elements made up of two bifilars are driven with equal amplitude signals and in phase quadrature. The four elements of the quadrifilar are fed with a 90° phase progression (0°, 90°, 180°, and 270°). The windings may be self-phased or externally phased.

In self-phasing quadrifilars, no external phasing network is used between the two bifilars. The quadrature phasing is achieved internally by using slightly different diameter bifilars. This is tantamount to using slightly different element lengths. Since the quadrifilar is a *resonant* structure, a bifilar with a diameter smaller than the resonant length will resonate above the desired resonant frequency. The slightly bigger elements will resonate below the design resonant frequency (see Davidoff and Stilwell). The smaller bifilar will act capacitively and the larger one inductively. This will produce leading and lagging currents (respectively) in the windings, which compositely produce a differential phase of 90°, satisfying the phasing between the two bifilar loops of the quadrifilar helix. A diagram of this structure is shown in Fig. 7.6. A single element is driven, and the remaining three are dummy elements which could be coaxial lines but are not used as such. The shields (if coaxial) of the bifilar are connected to the shields of the other bifilar as shown.

External phasing is produced by phasing networks as illustrated in Fig. 7.7. Both 0 to 180° hybrids and quadrature hybrids are used. The 0 to 180° hybrid can be a hybrid ring. The quadrature hybrids produce two equal outputs which are in phase quadrature. Both the hybrid ring and quad-

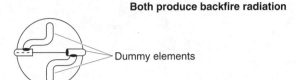

Figure 7.6
Self-phased and externally fed quadrifilar helices.

Both produce backfire radiation

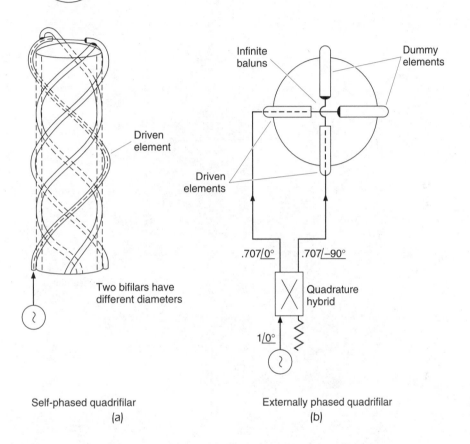

Self-phased quadrifilar
(a)

Externally phased quadrifilar
(b)

rature hybrids are illustrated in Fig. 7.8 with their associated scattering matrices. These devices are matched reciprocal devices (reflection coefficients = 0, or VSWR = 1). The diagonal elements in the matrix are therefore equal to 0.

In addition to the phasing network, the helices require a balun (balance to unbalance network) since the helices (balanced line) are normally driven by a coaxial line (unbalanced line). Baluns which may be used include the folded balun, split-sheath balun, or the infinite balun.

Another quadrifilar proposed by Kilgus using infinite baluns with two driven elements is shown in Fig. 7.6*b*. Two infinite baluns are

Figure 7.7
Bottom-fed quadrifi-
lar helix with external
phasing network.

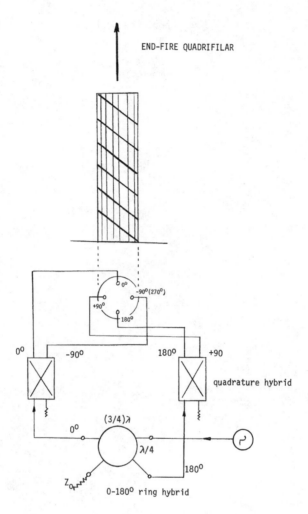

used, and external phasing of the two bifilars use a quadrature hybrid. Here, both bifilars are the same diameter in contrast to the self-phased structure.

One simple form of a bifilar using an infinite balun is shown in Fig. 7.9 on a bifilar winding set, so-called since it will work over extreme bandwidths, physical design permitting. One of the elements is a coaxial transmission line serving as the driven element with the other winding serving as the dummy element. (It may also be a coaxial line but the center conductor is not used.) The center conductor of the driven winding is connected to the end (or shield) of the dummy element as shown. This produces a backfire antenna beam, and no phasing network is required for a single bifilar helix.

■■ ■■ ■■ ■■

Figure 7.8
0 to 180° hybrid ring and quadrature hybrid junctions with their attendant scattering matrices.

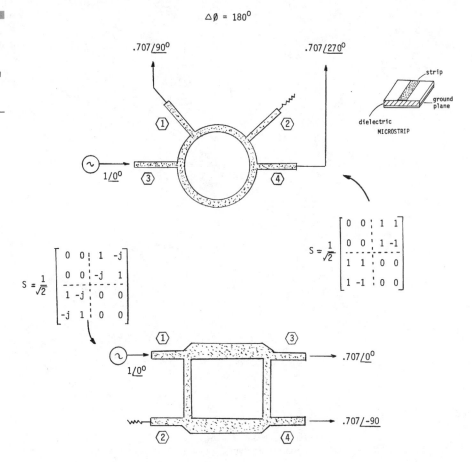

The radiation pattern from the helices can vary from an axial beam to a broad omnidirectional beam (conical beam) with greater radiation away from on-axis. The pattern directionality is controlled by the winding pitch and/or the diameter of the helix. For example, decreasing the number of turns (increasing the pitch) increases the beam width.

Examples of pattern variations are displayed in Fig. 7.10. For on-satellite applications with the beam directed at the earth, the shaped-beam patterns indicate that the gain is greatest near the horizon and has lower gain at the center of the beam, where the satellite is closest to the earth. Therefore, the radiation on the earth is more nearly uniform no matter where the earth station is in relation to the satellite. In more technical terms, the power-flux density incident on the earth in watts per meter square is uniform or isoflux. As indicated previously and also in Kilgus' paper, the direction of radiation in the elevation plane can be controlled by chang-

Figure 7.9
Backfire bifilar helix
antenna.

Backfire radiation

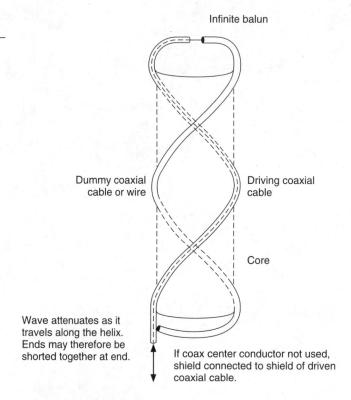

Infinite balun

Dummy coaxial
cable or wire

Driving coaxial
cable

Core

Wave attenuates as it
travels along the helix.
Ends may therefore be
shorted together at end.

If coax center conductor not used,
shield connected to shield of driven
coaxial cable.

• No ground plane required.
• No quadrature hybrid required.

ing the helix pitch angle and/or the helix diameter. The beam's omnidi-
rectionality can vary in elevation from zenith and/or nadir for the earth
station and satellite, respectively, to off boresight by increasing either of
the two parameters.

All the patterns shown in Fig. 7.10 are circularly polarized, and only the
top pattern shows a small axial ratio by the fine undulations on the pattern. That is, the pattern is practically circularly polarized in all θ (elevation) and ϕ (azimuth) directions. For perfect circular polarization, the
pattern would be smooth (axial ratio = 1 or 0 dB). Generally the axial ratio
deteriorates as the pattern is measured off boresight. Axial ratios in the
order of 1 to 2 dB are a good pattern. In antenna measurements, it is common practice to determine departure from circularity by illuminating
the test antenna with a rotating linearly polarized dipole or a linearly

Figure 7.10
Patterns resulting from measurements on several configured quadrifilar helices.

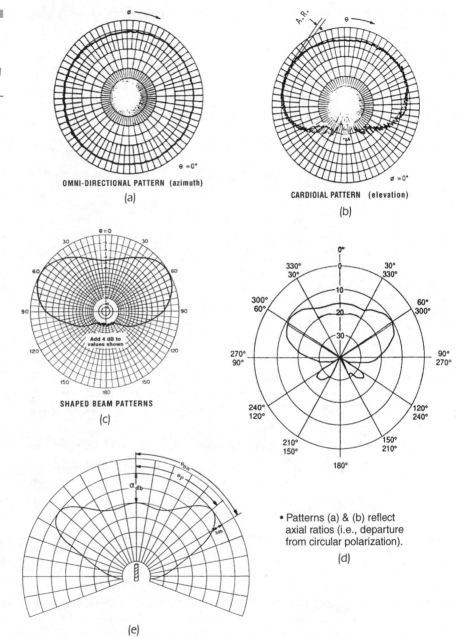

OMNI-DIRECTIONAL PATTERN (azimuth)
(a)

CARDIOIAL PATTERN (elevation)
(b)

SHAPED BEAM PATTERNS
(c)

• Patterns (a) & (b) reflect axial ratios (i.e., departure from circular polarization).
(d)

(e)

polarized waveguide horn. The other smooth trace patterns shown in Fig. 7.10 were measured with a fixed circularly polarized antenna with the same polarization as the tested antenna.

7.3.2 Microstrip Antennas

Microstrip antennas are also being considered for use in mobile-satellite service ground transceivers. *Microstrip* is basically a metallic patch supported by a dielectric substrate over a ground plane. It is compact, has a low profile, and can be designed to produce a wide broadside beam or a conical (omnidirectional) beam with circular polarization. A sketch of this antenna type is depicted in Fig. 7.11a.

The fundamental EM mode in microstrip is the TM_{11}. For excitation of this mode, circular polarization is achieved by driving the patch with two probes which are spatially separated by 90° and driven in phase quadrature. Excitation in this mode produces an axial beam with a near cardioidal pattern as shown in Fig. 7.11a. However, this pattern is not quite suitable for LEO mobile-satellite operation since, for angles off boresight, the gain is reduced, which is clearly not favorable for satellites at low-elevation angles.

It is possible to make the pattern more conical, with higher gain off boresight, by exciting the higher modes in the microstrip. This can be done by introducing additional probes as suggested by Huang. Some of the patterns calculated by exciting different modes are indicated in Fig. 7.12, which was taken from Huang's paper. Note that the peaks occur at different angles for the various modes. An experimental validation for one of the patterns produced by one of these modes is indicated in Fig. 7.13. The pattern manifests reduced gain at boresight corresponding to the zenith location if used in a ground transceiver.

A photograph of a microstrip antenna excited with a higher order mode is indicated in Fig. 7.14. If the reader looks closely, one can detect four probes. Actually, two are to excite the TM_{21} mode (note geometrically separated by less than 90°, but driven in phase quadrature), and the diametrically opposite two are mode suppressors.

Simple microstrip patches are relatively narrow band (2 to 3 percent) and are not operational over both the uplink and downlink bands of mobile-satellite systems. For example, for systems using 1.5 and 1.6 GHz the spread is 7 percent. Attempts have been made to increase the operating bandwidth by increasing the dielectric thickness. This will increase

Figure 7.11

(*a*) Microstrip patch producing circular polarization, and (*b*) broadband microstrip patch(es) for multi-band operation.

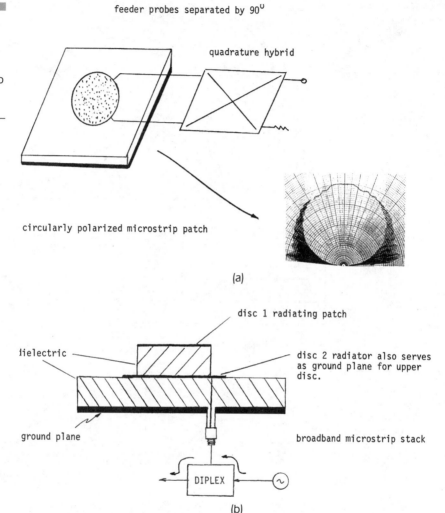

feeder probes separated by 90°

quadrature hybrid

circularly polarized microstrip patch

(a)

disc 1 radiating patch

dielectric

disc 2 radiator also serves as ground plane for upper disc.

ground plane

broadband microstrip stack

DIPLEX

(b)

the height profile, introduce multimoding, and will support surface waves having a deleterious effect on the antenna pattern and efficiency.

One method, which has been proven successful, is to stack one microstrip patch above the other. Figure 7.11*b* shows this stacking. The two discs resonate at different frequencies. For example, in LEO applications, they would resonate at 1.6 and 2.5 GHz. Dual band operation is also useful in the GPS navigational system which transmits at two frequencies: L_1: 1575.42 MHz and L_2: 1227.6 MHz. The patches are driven by a coaxial line

Figure 7.12
Calculated radiation
patterns of a higher
order mode in circu-
larly polarized circular
microstrip antennas.

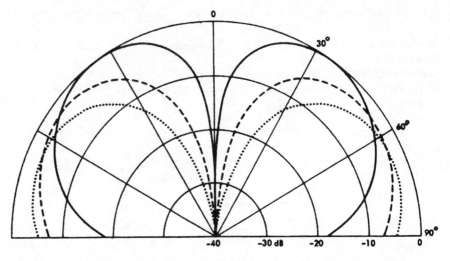

	MODE	RELATIVE DIELECTRIC CONSTANT	PEAK DIRECTIVITY	PEAK DIRECTION FROM ZENITH	RADIATOR DIAMETER
————	TM_{21}	1.25	6.9 dBi	35°	$0.9\lambda_0$
– – – – –	TM_{31}	2.2	4.6	54°	0.93
··············	TM_{41}	4.2	4.0	69°	0.88

Figure 7.13
Conical pattern gen-
erated by excitation
of a higher mode in
microstrip antenna.

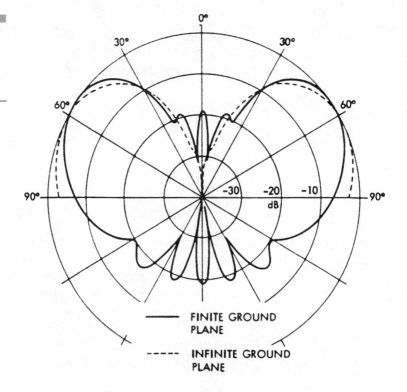

FINITE GROUND
PLANE

– – – – INFINITE GROUND
PLANE

Figure 7.14
An S-band microstrip
disk antenna using
higher order mode
(TM$_{21}$) to generate a
conical beam.

• Note probes locations.

passing through a clearance hole in the lower disc and attaching it to the upper disc. If the resonant frequencies are sufficiently separated, no diplexer is required. In this configuration, the bandwidth is *not* flat across the band between the two resonant frequencies, but the antenna operates at dual frequencies. Note that in mobile satellite operation, the uplink and downlink bands are separated, and continuous operation between the bands is not required or even desired. For example, for the band 1.5 to 1.6 GHz, continuous bandwidth would be 7 percent. The patches can therefore be made to resonate at smaller bandwidths around the operating bands. These bands are therefore about 1 percent for an up or down bandwidth of about 15 MHz. Patch designs are also available which can produce single-feed circular polarization and do not require two feeds as shown in Fig. 7.11*a*.

Microstrip patches have also been used in antenna arrays to produce medium-gain antennas with directional antenna beams. These will be briefly discussed in the section on antenna arrays.

7.3.3 Drooping Dipole

Drooping dipole low-gain antennas are also being considered for mobile satellite applications for LEO satellites (as well as GSO satellites). This is illustrated in Fig. 7.15. Shown is the inverted V-bend configuration, but other shapes include the inverted U. The dipoles are drooped downward to increase circular polarization at the low-elevation angles. The distance from the ground plane to the dipoles affects the pattern shape. That is, it can produce a beam which can go from an axial beam or cardioidal-type pattern on-axis to a conical beam with omnidirectional radiation. The

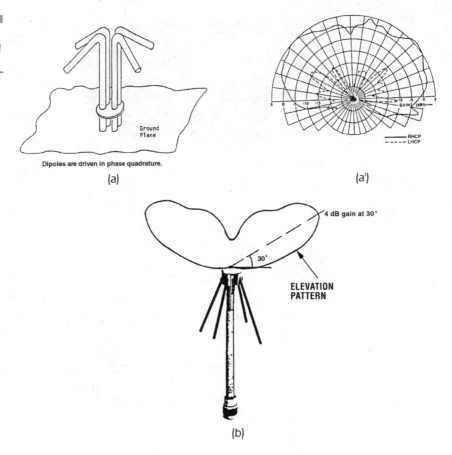

Figure 7.15
Drooping orthogonal
dipole antennas.

Ground
Plane

Dipoles are driven in phase quadrature.

(a)

RHCP
LHCP

(a')

4 dB gain at 30°

30°

ELEVATION
PATTERN

(b)

pattern (part *b*) in Fig. 7.15 results from raising the dipoles higher above the ground plane than the antenna whose pattern is shown in part *a* of the figure. Figure 7.15 also illustrates right-hand and left-hand circular polarization from the antenna on the upper left. Clearly, the antenna is designed for right-hand circular polarization.

To achieve the quarter wave relationship between the electric vectors produced by the dipoles, the space orthogonal dipoles are driven by a quadrature hybrid. Circular polarization may also be realized by self-phasing using dipole sets of different lengths. The shorter dipole acts capacitively (leading phase), and the longer dipole acts inductively (lagging phase), resulting in producing circular polarization. In this case, no costly quadrature hybrid is required.

7.3.4 Flat Spiral Antenna

Since being invented by Edwin Turner in 1955, the spiral antenna has found numerous applications in radar, electronic warfare, and, in recent years, in communication systems. They have been used singularly or as antenna elements in array systems. The antennas radiate circularly polarized waves, and the sense of polarization is dictated by the winding sense.

Flat spirals have evolved into two basic configurations: the so-called log equiangular spiral and the arithmetic or Archimedes spiral. These are depicted in Fig. 7.16. The Archimedean spiral also has a rectangular counterpart. Normally, the spirals are backed with a cavity to prevent radiation

Figure 7.16

Two forms of flat spirals.

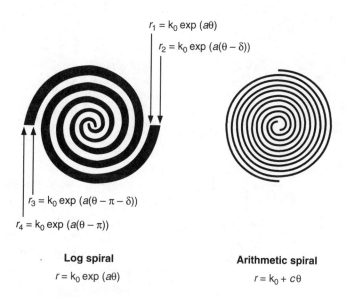

$r_1 = k_0 \exp(a\theta)$

$r_2 = k_0 \exp(a(\theta - \delta))$

$r_3 = k_0 \exp(a(\theta - \pi - \delta))$

$r_4 = k_0 \exp(a(\theta - \pi))$

Log spiral

$r = k_0 \exp(a\theta)$

Arithmetic spiral

$r = k_0 + c\theta$

a = Determine rate of growth

k_0 = Size of terminal region (radius at the start)

δ = Determines arm width

θ = Angle of rotation

- For the log spiral, the angle between the radius vector and the spiral remains the same for all points on the curve.

- A line drawn from the center outward; the spacings from one conductor to the next one is a constant ratio.

Geometry in feed region determines highest operational frequency.

Geometry in truncated region determines lowest frequency of operation.

Both spirals shown emanate right hand circular polarization out of page.

in the back direction. Most spirals consist of two arms, since their performance is more easily predictable and because the arms can be balun fed and perform better than single-arm units.

The equations for the two spirals shown in Fig. 7.16 are given by

$$r(\theta) = k_0 \exp(a\theta) \qquad [7.6]$$

for the logarithmic spiral and

$$r(\theta) = k_0 + a\theta \qquad [7.7]$$

for the Achimedean spiral.

$$r(\theta) = k_0 + a\theta \qquad [7.8]$$

where r = radius of spiral
k_0 = radius at the terminal region
a = determined rate of growth
θ = angle of rotation

For the spirals shown, radiation will be right-hand circularly polarized coming out of the page. If the structures are not backed (cavity) there would be bidirectional radiation with the radiation into the page being left-hand circularly polarized.

Spirals can be made to operate in two different modes of operation: (1) the axial mode, which produces a rotationally symmetric beam on the antenna axis, or (2) a normal mode which produces an omnidirectional beam or split-beam mode. For the axial mode, the terminals are driven in conjugate. In the normal mode, the terminals are driven in phase. An example of an axial beam pattern for an Archimedean spiral is shown in Fig. 7.17. To produce the axial mode, the antenna turns are driven in antiphase, and ideally this could be done with a balanced line* (like twin lead). Balance drive prevents anomalous behavior of the spiral. Practically, the antenna is driven by an unbalanced coaxial line with a balun transformer. Baluns for broadband antenna systems can be quite a design art. For the normal mode, the terminals are driven in phase, and a balun is not required.

It is of interest in some systems to generate both axial and normal beams. This can be achieved by driving the spiral terminals as shown in Fig. 7.18. As shown, a hybrid ring is used to generate both in-phase and

---

* A balanced two-conductor transmission line has equal currents of opposite phase in the line conductors at any given cross section.

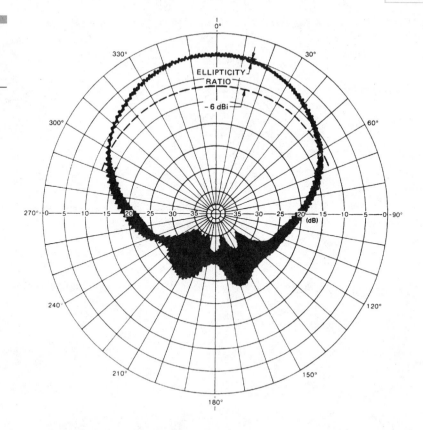

Figure 7.17
Typical pattern of an Archimedean spiral antenna.

antiphase current drives which in turn generate the two beams. In addition, a conjugate hybrid junction serves as a good balun, and some can be quite broadband.

7.3.5 Array Antennas

Arrays are candidates for mobile satellite applications. Research in this area is being performed in the United States, Canada, and Japan. There are basically two types which are being considered: mechanical scanning beam arrays and electronic beam scanning. The array's beam can be directional in elevation and azimuth. Because of their application to vehicle or aircraft deployment, their size is restricted to 1 or 2 ft linearly (if used as a linear array or in diameter for a planar array). A horizontal linear array will have a narrow beam in azimuth and a wider beam in elevation. The wider beam width in elevation does away with mechanical scanning in this direction.

Figure 7.18
Simultaneous genera-
tion of normal and
axial modes.

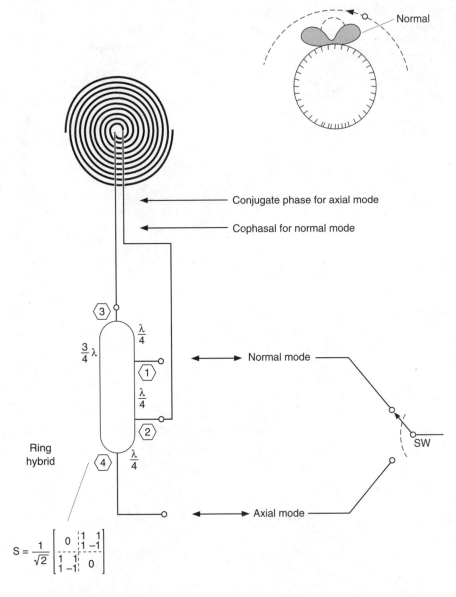

Most of these arrays have medium gain in the range of 10 to 15 dB, in contrast to the omnidirectional antennas, giving hemispherical coverage and having gains in the order of 2 to 5 dB.

An example of a mechanically steered linear array is shown in Fig. 7.19. The array consists of eight microstrip patches which produce a wide beam in elevation and a narrow beam in azimuth. The array is further divided in half for satellite tracking using a phase monopulse sensor with sum and difference output signals from the monopulse comparator. After

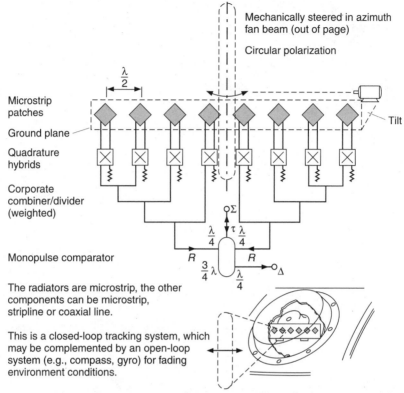

Figure 7.19
Linear array with RF monopulse sensing for satellite tracking.

Mechanically steered in azimuth fan beam (out of page)

Circular polarization

Microstrip patches

Ground plane

Quadrature hybrids

Corporate combiner/divider (weighted)

Monopulse comparator

The radiators are microstrip, the other components can be microstrip, stripline or coaxial line.

This is a closed-loop tracking system, which may be complemented by an open-loop system (e.g., compass, gyro) for fading environment conditions.

Mechanically steered linear array with a weighted aperture, and azimuth plane direction sensing through the vehicle of monopulse. In the transmit mode, the sum port is driven.

the array is steered in azimuth to the satellite bearing (sensed by the sum port output), the array goes into the tracking mode using monopulse. The two phase centers of the two halves of the array are located at the center of each subarray as shown in Fig. 7.20. The relationships at the output ports of the comparator are indicated on the figure. The signals undergo further processing to boresight the beam at the satellite location.

For a linear array using phase shifters in each element, the satellite is acquired by an azimuth search. After acquisition, tracking can be performed by dithering the antenna lobe about the approximate satellite position as shown in Fig. 7.21. This dithering of the beam produces unequal signals in the two lobe positions if the array is not boresighted on the satellite. In both the arrays described previously, the output of the error sensor is used to drive a platform driven by a motor to maintain boresighting of the array on the satellite azimuth.

Figure 7.20
Signal relationships in
linear array's phase
monopulse.

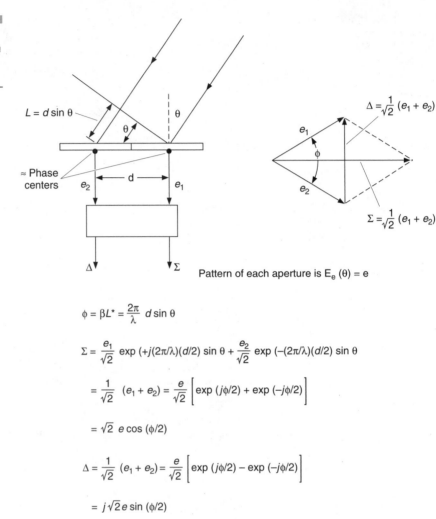

$$\phi = \beta L^* = \frac{2\pi}{\lambda} \, d \sin \theta$$

$$\Sigma = \frac{e_1}{\sqrt{2}} \, \exp \, (+j(2\pi/\lambda)(d/2) \sin \theta + \frac{e_2}{\sqrt{2}} \, \exp \, (-(2\pi/\lambda)(d/2) \sin \theta$$

$$= \frac{1}{\sqrt{2}} \, (e_1 + e_2) = \frac{e}{\sqrt{2}} \left[\exp \, (j\phi/2) + \exp \, (-j\phi/2) \right]$$

$$= \sqrt{2} \, e \cos \, (\phi/2)$$

$$\Delta = \frac{1}{\sqrt{2}} \, (e_1 + e_2) = \frac{e}{\sqrt{2}} \left[\exp \, (j\phi/2) - \exp \, (-j\phi/2) \right]$$

$$= j\sqrt{2} \, e \sin \, (\phi/2)$$

$*\beta = 2\pi/\lambda$, the propagation constant.

Steerable platforms are commercially available. A typical unit available from Kelvin Microwave is shown in Fig. 7.22. RF transmit and receive power is routed through the platform via the rotary joint and control power via the slip rings.

In the arrays described earlier, a signal from the satellite is required to aid in the acquisition and tracking. This is closed-loop tracking. In the real world where signal fading does exist due to obstacles such as trees and buildings, tracking loss can occur. Both systems must therefore use some

Figure 7.21
Hybrid open-
loop/closed-loop
satellite tracking
subsystem.

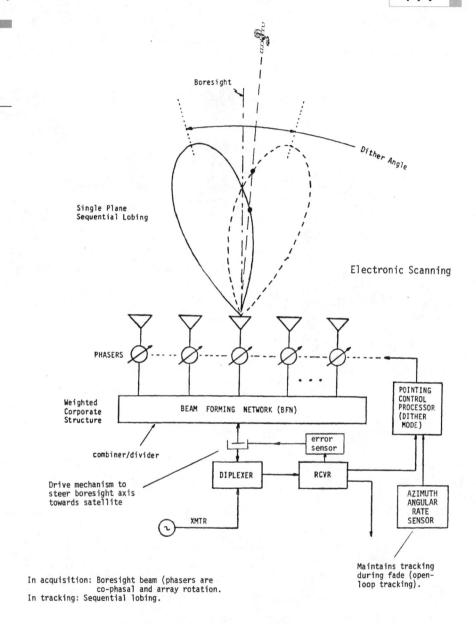

form of open-loop tracking in order to maintain pointing during the tracking phase (when the received signal fades below a prescribed value).

An observation made by jet propulsion laboratory (JPL) of unfaded and faded signals received from a satellite is shown in Fig. 7.23. For the unfaded case, the satellite signal is used for tracking. During heavy fade,

Figure 7.22
360° steerable
antenna platform.

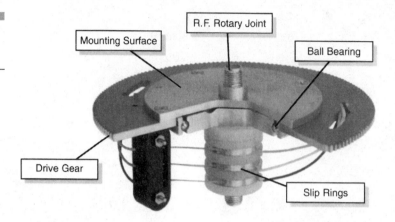

the satellite signal is unusable, and the tracking loop reverts to open-loop tracking. Sensors in this mode may include both geomagnetic sensors and gyros. The former can be affected by magnetic environments such as bridges or the vehicles themselves. Gyros are subjected to drift.

There are several attributes of open-loop and closed-loop satellite tracking which are of interest.

- *Open-loop satellite tracking.* No satellite signal required
  - Compass
  - Gyro
  - Navigational aids

Figure 7.23
A hybrid open- and
closed-loop pointing
system.

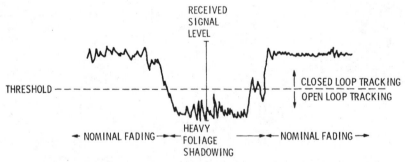

- DURING NOMINAL SIGNAL CONDITIONS THE CLOSED LOOP TRACKS THE SATELLITE SIGNAL AND CORRECTS FOR DRIFT IN THE OPEN LOOP

- DURING DEEP FADES THE OPEN LOOP MAINTAINS ACCURATE ANTENNA POINT

Debit side. Magnetic north variations across the United States, gyro drift, and inoperative navigational aids

Palliative. Revert to closed-loop tracking

■ *Closed-loop satellite tracking.* Depends on satellite signal

   ■ Monopulse

   ■ Sequential lobing

Debit side. Since depends on satellite signal, fading will reduce tracking accuracy

Palliative. Revert to open-loop tracking

■ *Optimum system.* Combination of the two previous tracking systems. During heavy fade, system goes into open-loop tracking. For normal signal conditions, closed-loop operation is used and corrects drift in open-loop operation.

7.3.6 Phased Array Antennas

Other more complicated (and costly) phased array antenna systems are being considered for mobile satellite applications. They possess rapid beam agility, low profile, and can provide beam directionality in two dimensions. On the debit side, they are more expensive because of the increase in the number of parts and are more labor-intensive. These antennas can steer their beams to wide angles off boresight but cannot provide full hemispheric coverage.

In the United States, Jet Propulsion Laboratory (under a NASA contract) has performed extensive research in this area. The array antennas were designed with moderate gains in the order of 10 to 15 dB and to be mounted on vehicles which can also include aircraft. Several subcontracts were awarded to antenna manufacturers for breadboards of planar arrays, and their work was monitored by JPL and performed under the MSAT-X program. Under contract were Bell Aerospace Corporation and Teledyne Ryan Electronics.

The overall system diagram of a typical phased array antenna subsystem is shown in Fig. 7.24. The arrays are designed to work at L-band (1.5 to 1.6 GHz). The array uses both closed-loop and open-loop tracking, and they were designed to use sequential lobing. Lobing is only a few degrees around the beam boresight position (function of the main beam skirt falloff). Open-loop tracking uses an azimuth-angle rate sensor.

Figure 7.24
Electronically steered
phased array
antenna, using
closed-loop and
open-loop tracking.

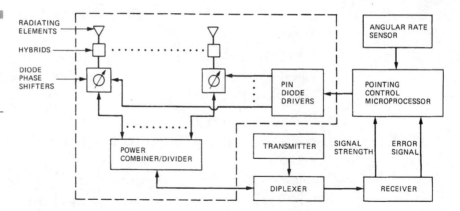

These contracts are now completed, and breadboards of the models have been delivered to JPL for further testing. The two arrays are depicted in Fig. 7.25. The performance goals which have been established for the arrays are indicated in Table 7.1.

A brief description of the respective designs and performance of the Bell and Teledyne array subsystems follows.

7.3.7 Planar Phased Array A

Array A structure is a planar circular array which is 21 in in diameter. The number of elements is 19 which are deployed in a triangular lattice in order to reduce the number required (see Jet Propulsion Laboratory). The separation of the elements is such that it prevents the advent of grating lobes into real space. The onset of grating lobes can be prevented if the following relationship is satisfied:

$$d' = \lambda_{\text{highest}}[1 - (1/N)]/1 + \sin \theta_s \qquad [7.9]$$

where d' = projected element separation in direction of scan
    θ_s = scan angle off boresight
    N = number of elements in direction of scan, $1/N$ factor included
        to prevent grating lobe skirt from entering real space

The geometry of the array and its element deployment is shown in Fig. 7.25. The antenna elements are cross-slot cavity backed, in stripline.

Figure 7.25
Geometries of 19 element planar phased arrays for mobile satellite service.

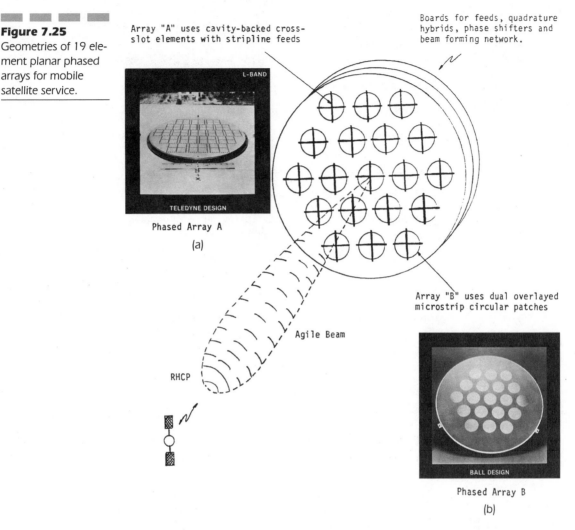

Array "A" uses cavity-backed cross-slot elements with stripline feeds

Boards for feeds, quadrature hybrids, phase shifters and beam forming network.

L-BAND

TELEDYNE DESIGN

Phased Array A

(a)

Array "B" uses dual overlayed microstrip circular patches

Agile Beam

RHCP

BALL DESIGN

Phased Array B

(b)

Each is driven by a four-probe network (see Fig. 7.26a) to produce circular polarization. The cavity is formed by plated holes from the top ground plane to the bottom ground plane and attached to the ground planes.

Each antenna element is driven by a phase shifter for steering the beam in elevation and azimuth. Eighteen phase shifters are used with the center element not requiring a phase shifter and serving as the reference. The 3-bit phase shifters are designed using switchable line lengths.

TABLE 7.1

Performance goals
established for the
arrays.

Frequency of operation	1545.0 to 1559.0 MHz Receive. 1646.5 to 1660.5 MHz Transmit. BW (each band): approximately 1%
Coverage	20° to 60° above the horizon, 360° in azimuth
Antenna gain	=>10 dB in the coverage area
Sidelobe level	<16 dB
Polarization	RHCP with maximum axial ratio of 4 dB across the scan range (30° to 70° off boresight)
Azimuth beam width	Approximately 25°
Elevation beam width	Approximately 35°
Satellite tracking	Azimuth tracking with occasional elevation search for maximum signal strength
Tracking technique	Sequential lobing augmented by azimuth angular rate sensor
Array size	Approximately 22-in diameter and 1-in height
Number and types of antenna elements	19, using microstrip or stripline
Phase shifter type	3-bit diode
Intersatellite isolation for orbital (GSO) reuse	20 dB for satellite separation of 35°

Line lengths are programmed in or out when scanning. This configuration is shown in Fig. 7.27a. Diodes are used to switch the line lengths in or out.

Switched-line shifters (also referred to as time-delay steering) have the desirable property that beam position is *independent of the frequency* driving the phaser. It is of interest to expound on this concept.

The phase shift $i\Delta\phi$ required to scan to an angle θ_s off boresight is given by:

$$i\Delta\phi = (2\pi d_i/\lambda) \sin \theta_s$$

$$= (2\pi d_i f/c) \sin \theta_s \qquad i = (n \mp 1), \quad n = \pm 1, \pm 2, \pm 3,\ldots \qquad \textbf{[7.10]}$$

where d_i = distance of the ith element from an arbitrary reference point
θ_s = scan angle
$\Delta\phi$ = phase shift between elements

Figure 7.26
Antenna elements
used in phased
arrays.

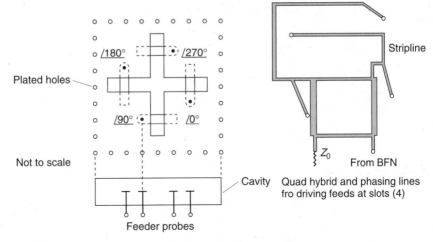

Plated holes

Not to scale

Cavity

Feeder probes

Stripline

Z_0

From BFN

Quad hybrid and phasing lines
fro driving feeds at slots (4)

Array A cavity-baked cross slot array element

(a)

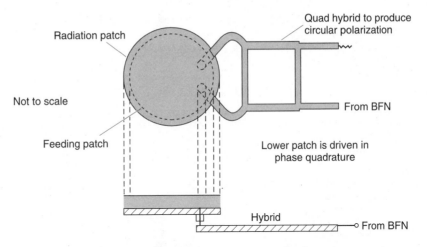

Radiation patch

Quad hybrid to produce
circular polarization

Not to scale

From BFN

Feeding patch

Lower patch is driven in
phase quadrature

Hybrid

From BFN

Array B dual resonant stacked microstrip elements

(b)

Note that if there is a frequency change (e.g., the transmit and receive signal frequencies are different), the phase and steering angle direction will change. Therefore, beam pointing errors occur where the transmit and receive beams do *not* point in the same direction.

If one uses a delay line as a phasing network, the phase shift is equal to $\Delta\phi = \beta L$ where $\beta = 2\pi/\lambda$. Plugging in Eq. [7.10] and solving for $\sin \theta_s$, we obtain

Figure 7.27
Phase shifter types
used in phased arrays
A and B.

Phased Array "A" Phase Shifter Type

Switched Line Length Phaser

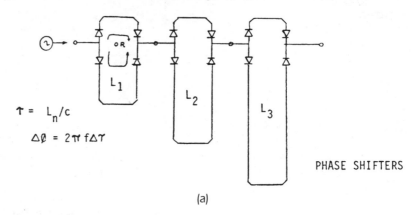

$\tau = L_n/c$

$\Delta \emptyset = 2\pi f \Delta \tau$

PHASE SHIFTERS

(a)

Phased Array "B" Phase Shifter Types

Loaded line Phaser   Reflection Phaser

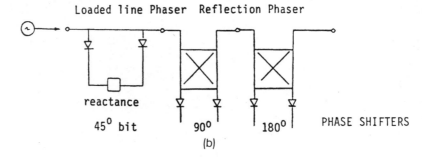

reactance

45° bit   90°   180°   PHASE SHIFTERS

(b)

$$\sin \theta_s = (c/2\pi df)\phi$$
$$= (c/2\pi df)\beta L$$
$$= (c/2\pi dg)(2\pi Lf/c)$$
$$= L/d \qquad \qquad [7.11]$$

where L = switched-in line length
   β = propagation constant, $2\pi/\lambda$

The steering direction is therefore independent of frequency. Since in mobile systems the transmit and receive frequencies are different, using delay lines for the phasers will not cause differential pointing of the transmit and receive beams.

It is clear from Eq. [7.10] that, since the scan angle should be the same for the transmit and receive frequencies, we can equate the two scan angles $\sin \theta_{sT} = \sin \theta_{SR}$. We therefore obtain:

$$\phi_T c / 2\pi df_T = \phi_R c / 2\pi df_R \qquad [7.12]$$

Assume that the *same* switched delay line length is plugged in for transmit and receive. The phase shift for transmit and receive is $2\pi L/\lambda_T$ and $2\pi L/\lambda_R$, respectively. We have

$$(2\pi L/\lambda_T)c / 2\pi df_T = (2\pi Lc/\lambda_R)c / 2\pi df_R \qquad [7.13]$$

$$L/d = L/d$$

This result indicates that the same phase shifter can be used for transmit and receive, and both beams will coincide. A 3-bit phase shifter using switched-in delay lines is shown in Fig. 7.27a.

Two pattern measurements made on array A are shown in Fig. 7.28. The broadside pattern is displayed in Fig. 7.28a, and Fig. 7.28b is the pattern steered 40° off boresight. Both measurements were taken at 1545 MHz.

7.3.8 Planar Phased Array B

Phased array B is also a planar phased array using 19 elements and a diameter of 24 in. The antenna elements are deployed in a triangular lattice. The antenna element is a dual-stacked disc in order to resonate at the two operating bands, 1.5 and 1.6 GHz. This element is shown in Fig. 7.26b. The lower disc is fed by two arms of a quadrature hybrid. Note that the probes are not separated by 90° since higher order mode excitation is used (see section on microstrip antennas). Right-hand polarization is used. The array geometry is shown in Fig. 7.25b.

The phase shifters use three bits with a design differing from that used in array A. The 45° bit is a loaded line, and the 90° and 180° bits are reflection-type phasers. The geometry of the phase shifter is indicated in Fig. 7.27b. The complete phase shifter requires six PIN diodes. Phased array A using switched lines requires 12 diodes.

The beam-forming network (BFN) is integrated into a single planar board with the phase shifters. The BFN is a corporate structure which is weighted radially across the aperture to improve the sidelobe level. Amplitude tapering is used. An example of amplitude tapering is shown in Fig. 7.29. This is not necessarily an optimum weighting of the aperture. The

Figure 7.28
Two radiation patterns for the teledyne phased array.

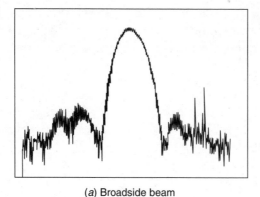

(*a*) Broadside beam

0   30   60

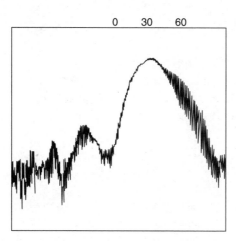

(*b*) Beam scanned 30° off boresight

tapering is accomplished by binary dividers, and these dividers may be the Wilkinson-type (requiring balancing resistors) or ring hybrids. (A better illumination function staircase would be 0, −3, −6, −9 dB.) Both can easily be designed in planar construction. Note that the center element in the array does not have a phase shifter. The beam-forming network is a corporate structure and is mounted on the same board as the phase shifters.

7.4 GSO Orbital Reuse

Amplitude tapering of the array's antenna illumination function reduces the sidelobe levels permitting frequency reuse and closer positioning of

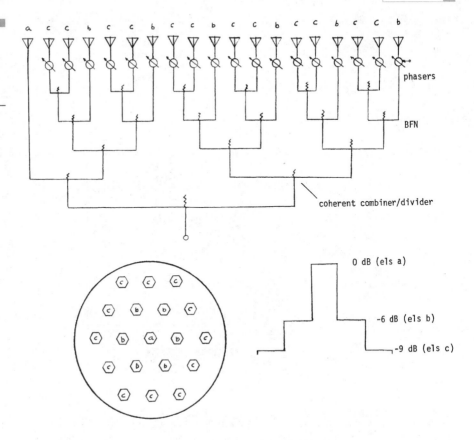

Figure 7.29
Weighted corporate beam-forming network used to reduce antenna sidelobes.

adjacent satellites. Table 7.1 indicates that the intersatellite isolation for orbital reuse should realize a carrier-to-interference ratio $C/I = 20$ dB for a satellite separation of 35°. This further assumes that the adjacent satellite will use the opposite sense circular polarization. For example, Fig. 7.30 shows that the desired satellite uses right-hand circular polarization and the adjacent satellite uses left-hand circular polarization. The isolation for a satellite separation indicated is 20 dB.

The azimuth pattern for one of the arrays described previously is shown in Fig. 7.31 (see communication from Dr. J. Huang). Both copolarization (co-pol) and cross-polarization (cross-pol) are depicted. The co-pol beam width is about 30°, and the on-axis gain of the co-pol is about 10 dB.

For a $C/I = 20$ dB criterion for orbital reuse, the satellite radiating the interfering signal can be located at a longitude distance of about 25° from the bona fide or desired satellite. This assumes that the interfering satellite uses opposite sense polarization signals, as indicated in Fig. 7.29.

Figure 7.30
Calculated pattern for
a medium-gain
phased array
antenna.

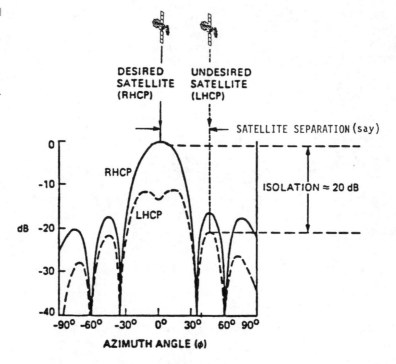

7.5 Beam Pointing System for Satellite Tracking

Both arrays A and B use a hybrid system of open-loop and closed-loop tracking to maintain the beam pointing at the satellite, which is very similar to that shown in Fig. 7.21. However, the beam is boresighted electronically and not mechanically. The array uses an angular rate sensor for open-loop tracking for conditions in which the signal is momentarily lost due to signal fading. When an adequate satellite signal level is present, closed-loop sequential lobing of the beam is used to zero in on the satellite location. This lobing is performd only in the azimuth plane.

A summary of the two phased-array parameters is given in Table 7.2.

Phased arrays are attractive for many applications because of their beam agility, multimode operation, and low profile. They also have some inherent disadvantages:

■ As the beam steers off boresight, the gain reduces, the beam distends, and asymmetry develops in the pattern.

Figure 7.31
Azimuth patterns measured at 20° elevation angle for a planar phased array.

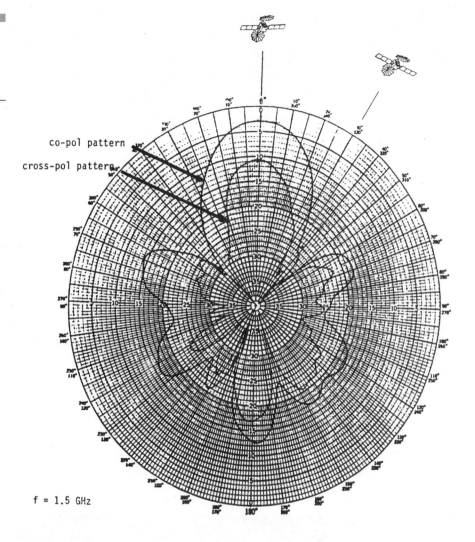

co-pol pattern

cross-pol pattern

f = 1.5 GHz

- Circular polarization degrades (increase in axial ratio) as the beam moves closer to the array surface, since the horizontal component does propagate well along the array.
- Steering beyond about 60° produces extreme deterioration of the antenna pattern.
- For an array of the same size, the mechanical has greater gain because the phased array has additional losses due to the phase shifters and other components.
- High cost occurs because of component count and labor-intensive upkeep.

TABLE 7.2

Summary of
Phased Array Pa-
rameters (Bread-
board Models)

Parameter	Phased array A	Phased array B
Array configuration	Planar circular	Planar circular
Frequency of operation	1545 to 1660.5 MHz	1545 to 1660.5 MHz
Diameter × thickness	21 in × 0.7 in	24 in × 1.3 in
Number of elements	19	19
Element type	Cavity backed; cross slots	Dual resonant; stacked circular; microstrip patches
Element beam width	≈140°	≈90°
Polarization	RHCP	RHCP
Axial ratio	≈8 dB @ 20° el	≈8 dB @ 20° el
az/el beam width	Approximately 25°/35°	Approximately 25°/35°
Antenna gain	#	#
Phaser types (18 els only)	Switched line lengths	Hybrid: loaded line, reflection
Number of phaser bits	3	3
Satellite tracking	Open-loop/closed-loop open-loop: angular rate sensor closed-loop: sequential lobing	Open-loop/closed-loop open-loop: angular rate sensor closed-loop: sequential lobing
Intersatellite Iso.	≈25 dB*	≈18 dB*

	f	*A*	*B*
Gain at 20° el.	1545 MHz	9.2 dBic	8.2 dBic
	1660	7.8	8.1
60° el.	1545	12.8	11.5
	1660	12.2	12.4

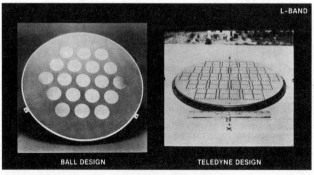

*Adjacent satellite is transmitting LHCP. In addition, the values are based on information received from
Dr. J. Huang of JPL, after the Mobile Satellite Conference (JPL, 1988).

7.6 Some Practical Array Tests in the Field

The first tests on an airborne phased array antenna were performed in Japan by the Communications Research Laboratory (see Taira et al.). The phased array was mounted on top of the fuselage of a Boeing 747 and used in flights from Japan to Anchorage, Alaska. The array communicated with the engineering test satellite V (ETS-V), a geostationary satellite located at 150°E longitude. The base station for the satellite link was located in Japan, and the base station to satellite link operated at C-band. The satellite to aircraft link was at L-band (1.5 to 1.6 GHz).

A photograph of the array mounted on the top of the Boeing aircraft is shown in Fig. 7.32. The 2-tier, 16-element array uses resonant frequency (1.5 to 1.6 GHz) circularly polarized microstrip patches, driven by 4-bit diode phase shifters. The full antenna consists of two identical arrays placed back-to-back, and one or the other is programmed into operation depending on the direction of the aircraft flight (east or west).

The array beam was capable of being steered ±60° in azimuth and about 26° in elevation. Being basically a linear array, its elevation beam width is wider than the azimuth beam width. Beam width in elevation makes the scanning less critical.

Radiation pattern measurements made at 1.5 GHz are illustrated in Fig. 7.32 for azimuth scanning up to ±60°. The patterns show the axial ratio and also indicate the degradation in this ratio as the beam is steered far off boresight. The gain falloff and beam distension, as the beam is steered off boresight, is in accordance with array theory.

Because phased arrays do not produce viable antenna beams when steered beyond about ±60°, there are areas of space which cannot be covered. This is demonstrated in Fig. 7.33 and indicated by the speckled areas. This has been referred to as the *keyhole void*. These volumes are predominantly in the fore and aft directions. Therefore, an aircraft flying in the direction of the satellite would not be seen by the array.

Attempts have been made to reduce these keyhole volumes by deploying small arrays in the fore and aft surfaces of the starboard and port arrays. Since these arrays will be smaller to maintain a small frontal profile, the beam patterns will be quite wide and need not be scanned. It has been claimed in an MSS application submitted to the FCC that more than 80 percent of the hemisphere can be covered by the starboard and port arrays and up to 95 percent if the fore and aft arrays are added.

Figure 7.32
TOYOCOM phased
array and experimen-
tal results.

(a) Phased Array Antenna Installed on the Top of a Boeing 747 Aircraft.

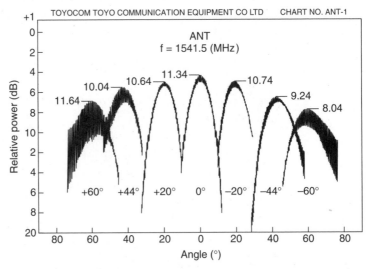

(b) Steered (=60°) radiation patterns for the TOYOCOM airborne phased array, showing the falloff in gain and increasing axial ratio as the beam is steered off broadside.

Figure 7.33
Keyhole voids (speck-led volumes) results from limitations on beam-pointing angles of phased array antennas.

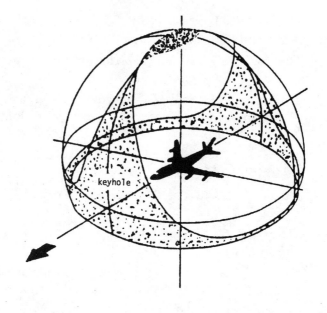

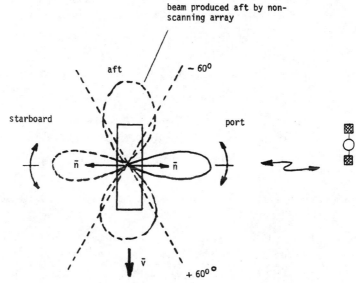

References

Adams, A., et al., "The Quadrifilar Helix Antenna," *IEEE Trans. Ant. and Prop.,* March 1974.

Bricker, R. W., and H. H. Rickert, "S-Band Resonant Quadrifilar Antenna for Satellite Communications," *RCA Engineer,* February/March 1975.

Chung, H. H., et al., "M-SAT-X Electronically Steered Phased Array Antenna System," *Proc. of Mobile Satellite Conference at JPL,* May 1988.

Davidoff, M., *The Satellite Experimenter's Handbook,* American Radio Relay League, Newington, Conn.

DuHamel, R. H., and G. G. Chadwich, "Frequency Independent Antennas," in *Antenna Engineering Handbook,* R. C. Johnson and H. Jasik (eds.), McGraw-Hill, New York, 1984.

Dyson, J. D., "The Equiangular Spiral Antennas," *IEEE Trans. Ant. and Prop.,* April 1979.

Gupta, K. C., "Recent Advances in Microstrip Antennas," *Microwave J.,* October 1984.

Hansen, R. C., *Microwave Scanning Antennas,* vol. ii, Academic Press, New York, 1966.

Hirata, T., "Development Status of Aeronautical Satellite Communication System at TOYOCOM Japan," *Space Communications,* July 1990.

Huang, J., "Circularly Polarized Conical Pattern From Circular Microstrip Antennas," *IEEE Trans. Ant. and Prop.,* September 1984.

Huang, J., Jet Propulsion Laboratory, Pasadena, Calif., personal communication with author.

Jet Propulsion Laboratory, MSAT Quarterly #7, October 1988.

Jet Propulsion Laboratory, several M-SAT-X quarterly reports.

JPL Technical Brief, "Low-Cost Omnidirectional Vehicle Antennas for Mobile Satellite Communications," November 1985.

Kaiser, J. A., "The Archimedean Two-Wire Spiral Antenna," *IEEE Trans. Ant. and Prop.,* May 1960.

Kilgus, C. C., "Resonant Quadrifilar Helix Design," *Microwave J.,* December 1970.

Kilgus, C. C., "Shaped-Conical Radiation Pattern Performance of Backfire Quadrifilar Helix," *IEEE Trans. Ant. and Prop.,* May 1975.

King, H. E., and J. L. Wong, "Helical Antenna," in *Antenna Engineering Handbook,* R. C. Johnson and H. Jasik (eds.), McGraw-Hill, New York, 1984.

Kraus, J. D., "The Helical Antenna," *Proc. IRE,* March 1949.

Kuhlman, E. A., and E. D. McKee, "Compact Antenna Has Symmetrical Radiation Pattern," NASA Tech. Briefs, Winter 1978.

Kumar, A., *Fixed and Mobile Terminal Antenna,* Artech House, Boston, Mass., 1991.

Long, S. A., and M. D. Walton, "A Dual-Frequency Stacked Circular-Disc Antenna," *IEEE Trans. Ant. and Prop.*, March 1979.

McCarrick, C. D., "A Vertically Mounted Elevation Steerable Vehicular Antenna for Mobile Satellite Communications," *Via Satellite*, October 1991.

MSS Application submitted to the FCC by American Mobile Satellite Consortium (AMSC), dated 1 February, 1988.

Ohmori, S., et al., "Aircraft Earth Station for Experimental Mobile Satellite System," *IEEE ICC*, Toronto, Canada, June 1986.

Pattan, B., "The Advent of Land Mobile Satellite Service Systems," *IEEE Trans. on Aerospace and Electronic Systems*, September 1987.

Ruperto, E. F., "The W3Kh Quadrifilar Helix Antenna," *QST*, August 1996.

Sharp, E. D., "A Triangular Arrangement of Planar Array Elements That Reduces the Number Needed," *IEEE Trans. Ant. and Prop.*, March 1961.

Shmit, F., "MSAT Mobile Electrically Steered Phased Array Antenna," *Proc. of Mobile Satellite Conference at JPL*, Pasadena, Calif., May 1988.

Stilwell, R. K., "Satellite Applications of the Bifilar Helix Antenna," *J. Hopkins APL Technical Digest*, no. 1, 1991.

Taira, S., et al., "High Gain Airborne Antenna for Satellite Communications," *IEEE Trans. Aerospace & Electronic Systems*, March 1991.

Transquilla, J. M., and S. R. Best, "A Study of the Quadrifilar Helix Antenna for Global Positioning System (GPS) Application," *IEEE Trans. Ant. and Prop.*, October 1990.

Weeks, W. L., *Antenna Engineering*, McGraw-Hill, New York, 1968.

CHAPTER 8

Operational and Proposed Global Mobile Satellite Systems

8.1 Introduction

Before the end of this millennium, satellite-based global communications will be an integral part of the wireless personal communications infrastructure. It was not too many years ago that mobile communication was solely the domain on Inmarsat, which provided voice and data services to ships at sea, and lately to aircraft and land. The terminals and the service calls are expensive.

With the advent of non-GSO telecommunications satellites, communication on a more personal level was possible, and in addition, the service blanketed the earth on a continuous basis. This will provide service in areas where there is marginal wireline infrastructure and may even likely be the only service. Satellite-based service to a region is a lot less expensive than laying down a wireline system. On the other hand, where there is an entrenched wireline service, landing rights may meet with some resistance.

At the present time, four systems are emerging to provide worldwide personal communication service, at least in the low earth orbit regime. They are Iridium, Globalstar, Odyssey, and IOC. At the GSO level, these include Garuda and Agrani.

Systems now in development and some nearing operational service will communicate with handheld transceivers, not unlike the present existing cellular phones. Since phone transmission standards are not universal, phones will be built to be compatible with standards of the countries which the satellites will overfly. Both GSM and CDMA accesses are battling for their global share.

Another area of development is the advent of systems providing cellular service from satellites in geostationary orbit. The technology of satellite advances has progressed to the point where large antennas in space are possible, and large solar arrays to generate the necessary power are in place. Large unfurlable antennas can provide high gain at even the lower microwave frequencies where these systems are operating (L-band and S-band). For the frequencies we are dealing with, antennas in the 20-m diameter range will be necessary. Solar array powers in the range of 10 to 15 kW are now feasible using efficient gallium arsenide solar chips.

8.2 Geostationary Altitude Mobile Satellite Service

8.2.1 Service to North America

The first use of the geostationary orbit to supply mobile satellite service was when Marisat went into operation. It began as a U.S. Navy program for ship-to-shore communications using the three UHF channels (on-board) and enabled the rise of the mobile satellite service industry. In 1996, a commemoration for its twentieth anniversary was held. In addition to UHF, there are also L-band channels using helical antennas which were leased by Comsat (Inmarsat) to provide commercial service. This was primarily for service at sea. These were followed by MARECS, IVMCS, and Inmarsat, but there are still two Marisats in operation (1996). The latest in this evolution is a constellation of Inmarsat-3 satellites which supply global service with global beams and spot beams. All of those previously listed have supplied service mostly for maritime use, but in a revision of charter, Inmarsat will now also supply service to aircraft. Communication service by the satellites mentioned previously is provided at L-band with the feeders in C-band. These systems do not provide service to handheld terminals. Comsat has developed a laptop sized terminal referred to as Planet 1 to be used with the Inmarsat-3 satellite. The terminal antenna is built into the cover. The satellite spot beams generate enough EIRP (and G/T) to communicate with the laptop terminal. For example, it has recently demonstrated this capability by placing a call from Malaysia to Germany.

In the continuing drama of mobile satellite communications, the FCC awarded a license in 1989 to a consortium of companies to supply service to the United States. The grantee is American Mobile Satellite Corporation (AMSC).* The satellite system has been dubbed AMSC-1. The service can be provided to what is generically referred to as the land mobile satellite service (LMSS), aircraft mobile satellite service (AMSS), and maritime mobile satellite service (MMSS). Voice, data, and facsimile service can be provided to portable units, vehicle terminals, or fixed sites. The service goes by the name Skycell. Cellular service (to handheld terminals) is not provided directly, but dual-mode operation is possible in areas where terrestrial cellular service also resides. In addition to its dual-mode satellite/

---

* Inmarsat is not permitted to serve the domestic market or North America.

cellular service, Skycell is also targeting trucking fleets and the maritime market. The AMSC satellite is not powerful enough to communicate with handhelds, because the earth station antenna must have gains in the order of 10 dB or better to achieve reliable voice service. The AMSC-1 satellite was launched in April 1995 and started to provide service several months later. AMSC-1 is located at 101°W longitude. The company has been authorized by the FCC to launch a total of three satellites.

Telesat Mobile, Inc. of Canada has a joint operating agreement to launch a comparable satellite (designated as MSAT), and the two will complement each other and also act as a replacement in case of failure of either satellite. The Canadian satellite has been launched and is located at 106°W longitude.

The number of regional spot beams produced by the AMSC-1 satellite is illustrated in Fig. 8.1. These include four spot beams which essentially

Figure 8.1
AMSC Skycell service to North America and coastal waters.

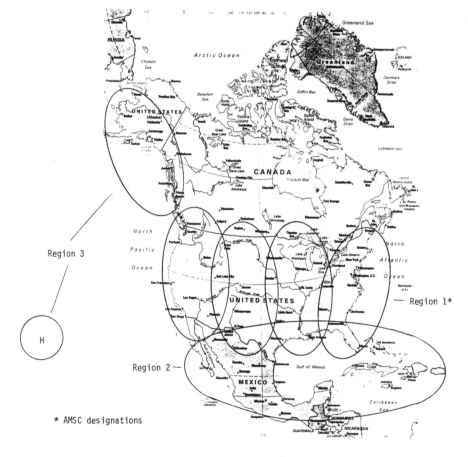

Region 3

H

Region 2 —

— Region 1*

* AMSC designations

cover the four time zones in the continental United States (CONUS). They are designated as Zone 1 through Zone 4, or in totality as Region 1 in the AMSC brochure. Other beams cover the southern coast of Alaska and its coastal waters. A small spot beam is placed on the Hawaiian Islands, and with the Alaska beam, they are referred to as Region 3. A third oblate shaped spot beam covers Mexico and the Caribbean and is referred to as Region 2.

The AMSC-1 satellite provides five kinds of service, but each does not employ all beams alluded to previously. This is illustrated in Table 8.1.

The frequencies of operation of the service links in the bands are 1530 to 1559 MHz on the downlink and 1631.5 to 1660 MHz on the uplink. This bandwidth is roughly enough to provide 2000 simultaneous duplex voice channels. The feeder links are in the 10-GHz band for transmission to the satellite and 13 GHz for the downlink. The satellite acts as a bent-pipe in which there is no onboard processing. The user terminals will use L-band. The routing of the signals to and from the satellite is shown in Fig. 8.2. Two unfurlable antennas are used for the communication service link. The SHF antenna is a dish which has a shaped beam that blankets most of North America.

There is no direct L-band to L-band communication between users (see Fig. 8.2). To make a call, the user uplinks a signal at L-band which is converted to 10 GHz in the satellite and downlinked at 13 GHz to a control center. The control center assigns channel pairs to the initiator of the call and the recipient. After this connection is made, communication between the two is possible. The initiated signal is uplinked to the satellite which downlinks it to a gateway which then retransmits it up to the satellite. Here it is converted to L-band and forwarded to its destination. If the receiver is not a mobile one, the gateway will connect the call to the local PSTN to complete the call. After the call is complete, the channel is returned to a pool. As indicated previously, this communication is basi-

TABLE 8.1	Service	Beams Utilized
Regional service provided by the various AMSC spot beams.	Land mobile service	CONUS beams, Caribbean
	Transportable service	CONUS beams, Alaska and Hawaii beams, Caribbean beam
	Fixed site service	CONUS beams, Alaska and Hawaii beams, Caribbean beam
	Maritime service	CONUS beams, Alaska and Hawaii beams, Caribbean beam
	Aeronautical service	CONUS beams, Alaska and Hawaii beams, Caribbean beam

Figure 8.2
AMSC spacecraft
dual-band payloads.

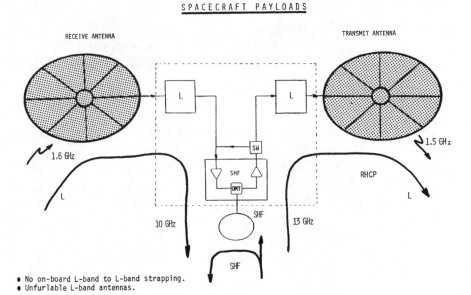

cally a double-hop system, and there is no direct L-band to L-band link. In technical terms, there is no L-band to L-band strapping in the satellite.

AMSC is the first to use dual-mode satellite/cellular terminals. If the mobile unit is not able to connect to the local terrestrial cellular system, the call is routed via the satellite mode.

8.2.2 Service to Europe Via Archimedes System

The European Space Agency has proposed that its member states fund development of a land mobile satellite system providing voice service from suitcase-size terminals. It was proposed to use a Molniya-type elliptical orbit with the apogee over Europe. The benefits of using this type of orbit are manifold. They provide much higher elevation angles of the earth station antenna beams and thus mitigate the fading which results from operating at low elevation angles due to multipath, blockage, and shadowing. Elevation angles hover in the range of 70°. If the satellite operated in geostationary orbit, the elevation angles would average to approximately 30°. In addition, the user antennas need not provide omnidirectional coverage since the satellite dwells for an appreciable amount of time in the apogee region. The antennas can be *fixed* directionally. These two factors (elevation angle and increased antenna gain due to directionality) permit a reduced margin in the link budget. Thus a considerable saving in satellite power is achieved.

The proposed satellite constellation is to use a four-satellite structure with each satellite in its own separate Molniya orbit, nodal crossings 90° apart, and inclined by the required 63.4° (to prevent apsidal rotation). The satellites are phased to blanket Europe around apogee at different times in order to achieve 24-h coverage. With an orbital period of 12 h, two apogees are produced in the northern hemisphere, but only the one over Europe will be activated. The apogee dwell is about 6 to 8 h, and in this interval, the satellites will be activated. The configuration for this orbital deployment is shown in Fig. 8.3.

In each satellite (for a period before it approaches and leaves apogee), the onboard antenna will illuminate Europe with six spot beams. Note that during this interval, the range to the earth station will change causing signal levels to change by about 4 dB. The footprint size will also change if not compensated by the satellite antenna, for example, by reducing the antenna spot beam width as the satellite approaches apogee. Clearly the reduced beam width will provide an increase in antenna gain where it is most needed since the slant range to the satellite is increasing. The system

Figure 8.3
Molniya-type orbits (4) used in the Archimedes mobile satellite system.

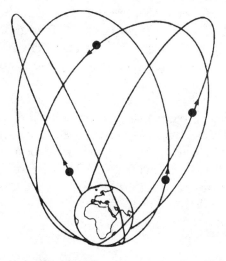

- Four satellites in their own individual Molniya orbits with nodal separations of 90 degrees.
- Each satellite reaches apogee region over Europe at different times of the day. Twntyfour hour coverage is maintained.

will provide service at L-band. The number of voice channels for each satellite will be about 3000.

The spacecraft configurations for the AMSC and Archimedes are similar. These are sketched in Fig. 8.4. Both use dual unfurlable mesh antennas which are 5 to 6 m in diameter. Both use offset feeds, but the AMSC design is a fixed labyrinth structure producing fixed beams. The Archimedes antenna uses a phased array feed which steers the six spot beams to maintain illumination of Europe, as each satellite moves toward the apogee region. Both use bent-pipe transponders (the uplink signal is converted to a lower downlink frequency and amplified). They can thus accommodate different access and modulation standards.

8.3 Non-GSO Mobile Satellite Service

In the evolution of mobile satellite service, the key is providing ubiquitous and personal communication to handheld transceivers at a reasonable cost. The advent of innovative signal processing and circuit miniaturization (MMIC, VLSI) has made this possible. A further refinement is interfacing with the existing terrestrial cellular infrastructure, thus making a dual satellite—terrestrial mode possible. In pursuit of this objective, the emergence of non-GSO orbit communication satellites came about. This includes satellites in orbits below 2000 km (LEOs) and satellites in medium altitude orbits (10,000 km). The United States has been the driving force behind this whole concept. In summary, these systems will supply personal communication voice and data services to handheld transceivers.

One issue which has impeded progress in this area includes the spectrum allocation and regulatory matters, both indigenously and internationally. However, the issue has been addressed by international bodies at the various WARCs and WRCs which have occurred since 1992. One momentous decision was made at the WARC-92 meeting to allocate the RDSS spectrum (1610 to 1626.5 MHz and 2483 to 2500 MHz) to LEO services on a worldwide basis. These frequency bands are allocated to the Big LEOs* which provide voice, data, faxing, and radiolocation determination

---

* In the United States the non-GSO satellites have been designated Big LEOs, which provide voice and datalike service. Little LEOs provide only datalike service. Big LEO service links operate above the 1-GHz band, and Little LEOs operate below 1 GHz (VHF/UHF).

Figure 8.4
Mobile satellite
spacecraft
configurations.

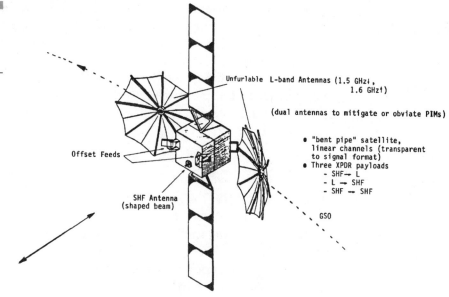

Unfurlable L-band Antennas (1.5 GHz↓,
1.6 GHz↑)

(dual antennas to mitigate or obviate PIMs)

Offset Feeds

- "bent pipe" satellite,
  linear channels (transparent
  to signal format)
- Three XPDR payloads
  - SHF→ L
  - L → SHF
  - SHF → SHF

SHF Antenna
(shaped beam)

GSO

(a) AMSC satellite in GSO orbit.

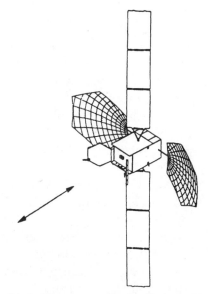

(b) Archimedes satellite in elliptical orbit.

services. In the United States, systems entitled Iridium, Globalstar, and Odyssey are in this category. The bands listed previously were segmented to allow system operation with different access techniques. These access techniques include FDMA, TDMA, and CDMA. There are some caveats in this band segmentation to allow other services to use small sections of the spectrum. This, for example, includes Radio Astronomy Service. This is discussed in more detail in Chap. 3.

8.3.1 Little LEO Mobile Satellite Service (First Round)

In the United States, the FCC has granted licenses to Little LEO systems operating below 1 GHz in the VHF/UHF bands. These operate in the store-and-forward mode providing data and messaging but no voice service. Generally these satellites are smaller and less complicated than the Big LEOs. Their operating altitudes are below 1300 km. They are also designed to operate with handheld transceivers. Liquid crystal alphanumeric displays can reveal the nonvoice information.

The three organizations granted licenses by the FCC are Orbital Sciences Corporation (ORBCOMM), Starsys Global Positioning System (Starsys), and Volunteers in Technical Assistance (VITA). ORBCOMM proposes to put up a 36-satellite constellation in four orbital planes inclined 45° with eight satellites in each plane. In addition, this configuration will include two planes with an inclination of 70° and two satellites per plane. ORBCOMM has petitioned the FCC for a modification to their system by requesting to deploy eight satellites per plane in their 70° inclined orbits. As of this date (January 1997), this petition is pending at the FCC.

Starsys will launch 24 satellites in six planes inclined 53°, with four satellites in each plane. VITA had attempted to launch one satellite into an 88° inclined orbit, but this was lost in a launch failure. The first two ORBCOMM satellites dubbed Microstar were launched in April 1995. Up to 36 are scheduled to be launched in 1997. The launch vehicle is their Pegasus XL which will be launched from the fuselage of a Lockheed L1011 aircraft. Pegasus XL can launch a cluster of eight satellites. The satellites are jettisoned at regular intervals in a single orbit. In order to ensure a uniform distribution of the satellites in the orbit, onboard cold gas thrusters will assist in the desired phasing of the satellites.

A listing of the salient parameters for the first round of licensed Little LEOs is indicated in Table 8.2.

TABLE 8.2

Salient Parameters for First Round Licensed Little LEOs (narrow band data, position determination, paging, messaging)

Constellation Parameters	ORBCOMM	Starsys	VITA
Service	Data, messaging, paging	Data, messaging, position determination	Technical assist to developing nations
Number of satellites	36 (later 48)*	24	1
Orbit inclination (all circular)	45°/70°	53°	88°
Altitude	785 km	1000 km	800 km
Number of planes/ satellites/inclination	4/8/45° 2/2/70°	6 planes 4 satellites per plane	1
Frequencies	137 to 138 MHz 148 to 149.9 400.15 to 401 ————————————————————————→		
Average time in view	10 min	15	10
Transponder type	Some processing, store and forward	Bent-pipe	Bent-pipe
Access	FDMA	CDMA	FDMA
Data rate, up/down	2.4/4.8 kbps	9.6 kbps	1.2, 9.6 kbps
Stabilization	Gravity gradient + torquers ————————————————————————→		

* Has petitioned the FCC to launch eight satellites each in two orbits with an inclination of 70°. This was submitted in second round small LEO applications. This is pending as of January 1997.

8.3.2 Second Round Little LEOs

In November 1994 the FCC allowed a second round of Little LEO applications to be submitted. Even though there is a paucity of spectrum for the second round LEOs, there may be enough to accommodate one (or possibly as many as three) in the designated bands. This may be achieved by dynamically utilizing the channels by time-sharing since the satellites are dormant during long intervals of their orbits. This is a relatively new approach (in satellite applications) at the FCC to obtain better utilization of the spectrum. In October 1996 the FCC issued a Notice of Proposed Rule Making (NPRM) proposing the assignment of frequencies to accommodate the licensing of three systems in the second round. Eight applicants have applied for licenses in the second round. However, three have also applied in the first round. The NPRM states that these three will not be considered in the second round. If adopted, the FCC's rules would deny additional spectrum to GE Starsys, ORBCOMM Global, and Volunteers in Technical Assistance. The reason for this exclusion is to enhance competition. The organizations in contention are CTA, Inc., E-Sat, Final Analysis, GE American Communications, Inc. (GE has bought into the first round Starsys, which could possibly exclude it in the second round), and Leo One USA. Essentially the notice indicates that time-sharing will be required and that Little LEO operators will be required to periodically shut off all ground terminals that are under the footprint of incumbent operator's satellites. Since there are five applicants vying for the spectrum, if they are all financially qualified and have mutually exclusive systems (cannot share the spectrum among themselves), the FCC proposes to hold spectrum auctions.

At the WRC-95 the United States was not able to obtain sufficient additional spectrum to accommodate the second round LEOs. However, it will attempt to obtain additional spectrum allocations at the WRC-97 Conference.

A summary of the parameters of the second round Little LEOs is indicated in Table 8.3. As of the date of this writing, all applications are on hold until the spectrum issue is resolved. The applicants marked with an X are first round permittees, too.

In January 1997, Final Analysis launched a small experimental LEO (FAISAT-1) on a Russian Cosmos rocket piggybacked onto a navigational satellite. As indicated previously, in April 1995, ORBCOMM launched two birds (first of 36) to become the first Little LEO to begin building an in-orbit satellite system aimed at providing global messaging services.

Figure 8.5 shows a montage of second round LEO configurations. All save LEO One USA use gravity gradient stabilization booms coupled with magnetic torquers. All booms are extended after the satellites reach their orbital altitude. For example, the length of the boom on the Microsat satellite is 8½ ft long. The differential gravitational force at opposite ends of the boom is really quite minute, but nevertheless is still sensitive enough to keep the boom oriented toward the earth. For this passive stabilization system, pointing can be maintained to within approximately ±5°. Gravity gradient systems will normally have one fixed nadir-pointing antenna beam with its axis colinear with the boom axis. The antenna used is usually a wire antenna such as a helix, cross Yagi-Uda, turnstile, or flat spiral. Note that these all produce circular polarization. LEO One USA will be three-axis stabilized. On Microstar, the VHF/UHF quadrifilar antennas are part of the stabilization boom.

TABLE 8.3
Second Round Little LEO Applicants

Constellation Parameters	CTA	E-Sat	Final Analysis	GE Americom	LEO One USA	X ORBCOMM	X Starsys	X VITA
Service	Store and forward	→						
Number of satellites	36	6	26	24	48	48		1
Planes/inclination	4/50° 1 sun-sync	2/100.7°	4/66°	4/98°	8/50°	4/45°		1/88°
Altitude	1000 km	1261 km	1000 km	800 km	950 km	775 km		800 km
Frequencies	138 MHz 148 MHz 400 MHz	→						
Access techniques	FDMA DCAAS	CDMA	FDMA DCAAS	FDMA DCAAS	FDMA DCAAS			FDMA DCAAS
Orbit life	5 yr	10 yr	7 yr	5 yr	5 yr	5 yr		
Satellite stabilization	Gravity gradient + torquers	→			3-axis	Gradient + torquers	→	

NOTE: CTA, E-Sat, Final Analysis, GE Americom and LEOSAT USA are new entries, where the others have previously submitted applications in the first round. GE Americom has bought into first round Starsys.

231

Figure 8.5
Montage of satellite
configurations for
second round LEOs.

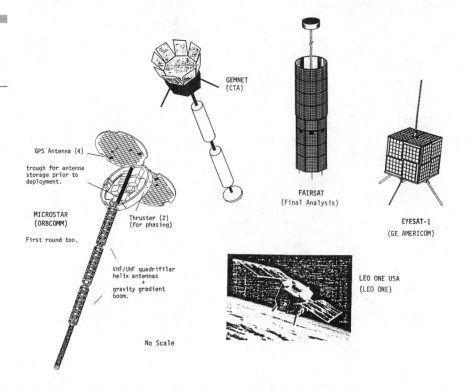

8.3.3 Voice and Data Big LEOs

In the early 1990s, six U.S. companies filed applications to provide satellite-based personal communications on a global and continuous basis. Five would operate at altitudes lower than the usual communications satellites operating in geostationary orbits. They have been dubbed non-GSO satellites and are designed to operate in low earth orbit (LEOs) and medium earth orbit (MEOs). The sixth applicant proposes to operate their system at a geostationary altitude.

The satellites operating at low altitudes will require a constellation of satellites in multiple orbits in order to provide continuous service, since each satellite will remain in view for only a small percentage of its time of orbit. Typically 10 to 15 min for LEOs and 2 h for MEOs.

The satellites are designed to provide voice, data, fax, and geolocation information. Service will be provided to handheld transceivers. The services are not unlike those provided by the present terrestrial cellular services, serving mostly urban areas where the market exists. The satellite-based service on the other hand can supply thin-route personal communications to remote areas in addition to ocean areas if necessary. The satellite system can therefore complement the terrestrial-based cellular systems and can operate in a dual mode. Actually, many of the satellite providers will design handsets for dual-mode operation and also for interfacing with the local telephone network in the areas being served.

In 1995 the FCC granted licenses to three of the applicants with the other two put on hold until they were able to show financial viability. The three included Motorola (Iridium), TRW (Odyssey) and Loral/Qualcomm (Globalstar). The proposed bands of operation for the systems are 1610 to 1626 MHz for the uplink operation and 2483 to 2500 MHz for the downlink. These bands are frequently referred to as L-band and S-band, respectively. The WARC-92 allocated these bands for satellite mobile service. Motorola chose to confine its operation for uplinking and downlinking to using only the L-band. The sections of the bands used by the three applicants have been discussed in Chap. 3.

The parameter comparison of several of the Big LEOs are shown in Table 8.4. Note that they all supply voice service in the band above 1 GHz. Also listed is the Intermediate Communication Global satellite constellation which is an offshoot of the Inmarsat organization. Globalstar, Iridium, and CCI-Aries are low earth orbit (LEO) proposals at altitudes below 1500 km. Odyssey and ICO Global are in medium altitude earth orbits (MEO) at an altitude of about 10,000 km. Ellipso-Ellipsat uses three orbits for their constellation. Two are in elliptical orbit with inclinations of 63.4° and 116.5° and an eccentricity of about 0.35. These are referred to as the borealis constellations. The third orbit is an equatorial (plane of the equator) circular orbit operating at an altitude of 7800 km. These orbits are shown next to the Ellipso satellite in Fig. 8.6. Both Ellipso and Aries have received licenses from the FCC.

The configurations for several of these satellites are depicted in Fig. 8.6. All are three-axis stabilized, and all use planar-array-type antenna for generating spot beams (cells) for user service at L-band. Iridium uses onboard processing and in addition uses cross-linking to transfer voice and data to other orbits in the constellation or to adjacent satellites in the same orbital plane. There are also Ka-band arrays and Ka-band horns; these are shown in Fig. 8.6 also. The arrays are for intersatellite communications, and the horns are for up-down communications to the gateways which interface with the PSTN. Globalstar and Ellipso will interface with the gateways via C-band frequencies.

A relatively new entry into this stable of constellations is ICO Global Communications. It is expected to provide personal communications service beginning in the next century. This system, as well as the competitive

TABLE 8.4

Comparison of Competing LEO Voice Satellite Systems

Constellation Parameters	Odyssey	Globalstar	Iridium	CCI-Aries	Ellipso-Ellipsat	ICO Global
Users/uses	Remote telephone, remote cellular, international travelers	Remote telephone, remote cellular, international travelers	Remote telephone, remote cellular, international travelers	Remote cellular, international travelers	International travelers with vehicle and fixed service needs	Remote telephone
Services	Voice, data, fax, paging	Voice, data, fax, RDSS, paging	Voice, data, fax, RDSS	Voice, data, RDSS	Voice, data, fax, paging, RDSS	Voice, data, fax
Coverage	Global	Global	Global	Global	Limited global (N of 50°S lat.)	Global
Orbital type	MEO	LEO	LEO	LEO	Elliptical-2; Circular-1	MEO
Altitude/period	10,354 km/<6 h	1414 km/114 min	785 km/100 min	1018 km/105 min	Apogee 7800+ km	10,354 km/6 h
Number of satellites (spares)	12 (2)	48 (8 spares)	66 (6 spares)	48	15 ellip.; 9 circ.	10—12 (2)
Planes/inclination	3/52°	8/52°	6/86.4°	4/90°	3/63.4°, 0°	2—3/45°
Satellite weight	1917 kg	426 kg	700 kg	<500 kg	625 kg and 650 kg	2000 kg?
Mission life	15 yr	7.5 yr	5 yr			10 yr
Frequency: Gateway Up/Down, GHz	19.4—19.6/29.1—25.25	5.091—5.250/6.875—7.055	19.3—19.6/29.1—29.4	C-band	C-band	5.179—5.2445/7.011—7.0739
User Up/Down, GHz	1.610—1.62135/2.4835—2.500	1.610—1.62135/2.4835—2.500	1.62135—1.6265	1.610—1.62135/2.4835—2.500	1.610—1.62135/2.4835—2.500	1.985—2.015/2.170—2.200
Transponder type	Bent-pipe	Bent-pipe	Processing			Bent-pipe

Spot beams per satellite	61	16	48	32	37	163
Number of FDX circuits per satellite	>3000	2800	2300			
Satellite cross-links, GHz	No	No	4/sat. @ 25 Mbps 23.18—23.38	No	No	No
Modulation	QPSK spread spectrum	QPSK spread spectrum	QPSK			QPSK
Multiaccess	CDMA	CDMA	TDMA	CDMA	CDMA	FDMA/TDMA
Satellite connect time	1—2 h	10—12 min	9 min			1—2 h
Minimum elevation	22°	10°—20°	8.2°			20°
Data rate (handset), kbps	4.8 (voice), 1.2—9.6 (data)	1.2—9.6 (voice) 2.4—9.6 (data)	4.8 (voice), 2.4 (data)			4800 MOS 3.5
FCC License	Jan-95	Jan-95	Jan-95	None	None	—
First launch date	1998	1997	1997	1997	1997	2000
Fully operational	1999	1998(4)	1998(4)			2000
Total system cost estimate	$2.5 billion	$1.9 billion	$4.7 billion	$1.7 billion	$800 million	$2.8 billion space seg.
Estimated per minute charge	$0.65 (wholesale)	$0.35—0.53 (wholesale)	$3.00 (retail) (retail)	$0.25—1.00 (wholesale)	$0.25—0.60 (wholesale)	$2/min
User terminal initial price	$700	$750	$2500	$1500	$700	$1000
Satellite antenna	Array	Array	Array	Array?	Array?	Array

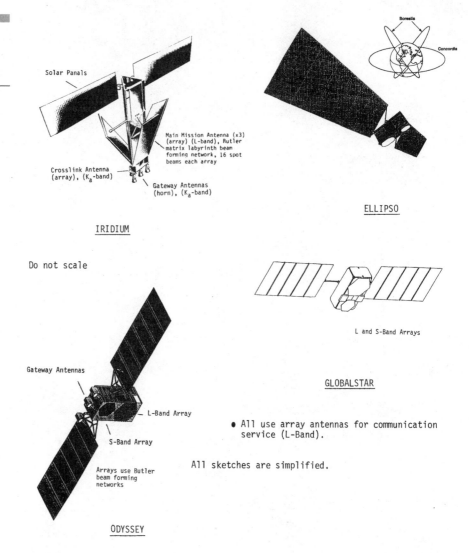

IRIDIUM

Solar Panels

Main Mission Antenna (x3) (array) (L-band), Butler matrix labyrinth beam forming network, 16 spot beams each array

Crosslink Antenna (array), (K$_a$-band)

Gateway Antennas (horn), (K$_a$-band)

ELLIPSO

Do not scale

GLOBALSTAR

L and S-Band Arrays

ODYSSEY

Gateway Antennas

L-Band Array

S-Band Array

Arrays use Butler beam forming networks

● All use array antennas for communication service (L-Band).

All sketches are simplified.

systems, will provide service to handsets as well as to fixed terminals. The handsets will operate in a dual mode, either with its parent satellites or with terrestrial PCS networks. Satellites for this system are presently being built by Hughes.

The first three Iridium satellites are expected to be launched in the early part of 1997. They will be lofted by a McDonald Douglas Delta II rocket from Vandenberg AFB in California. Actually, the Delta II can accommodate five Iridium satellites on a single launch. Other launch vehicles will be used which include the Russian Proton rocket which can manifest seven satellites and the Chinese Long March 2C/SD which can deploy satellites in pairs. The full complement of 66 satellites (plus six spares) will be in orbit by 1998. The 12-satellite TRW constellation expects to be fully operational in the year 2001, and the 48-satellite Globalstar constellation is expected to be fully operational before the turn of the century.

8.4 Geostationary Mobile Cellular Service

8.4.1 Regional Systems

The next generation of mobile satellite systems will provide personal communications service from the geostationary orbit. Several regional cellular systems providing service to handhelds are now in the proposal stage or are being built. In particular, these are being designed to cover regions on the Pacific Rim, Near East, and Africa. Another indigenous system being proposed in the United States is dubbed CELSAT, for which an application has been received by the Federal Communications Commission. It is presently awaiting a license.

Regional mobile satellite communications have several advantages:

- Instant capacity over large geographic areas
- Lower infrastructure costs
- Complementary with fixed wireline service and cellular infrastructure
- Direct call relay through satellite (single-hop mobile at L-band)

Systems presently being implemented and designed to cover several swaths from Africa to the Philippines include the following:

- *ACeS.* Asia Cellular Satellite System, which also has been referred to as Garuda. Possible orbital locations are at 118°E and 80.5°E. (One will be deployed initially.)
- *APMT.* Asia Pacific Mobile Telecomm
- *ASC.* Afro-Asian Satellite Communications. System referred to as AGRANI 1, and when deployed will be located at 21°E. Other positions and satellites are contemplated.

The ACeS will provide coverage to most of Asia from Pakistan to Japan including southern China and all Southeast Asian nations. APMT will cover a similar area and all of China. ASC will cover areas from Turkey to Singapore and parts of east Africa and the Middle East. In all, about 50 countries will be served. Several satellites will be used for this coverage. It is reported that ACeS and ASC will be compatible with the Groupe Speciale Mobile (GSM) standard.

Another proposed system is Satphone, in which Lockheed Martin International Limited is involved, and it would serve the Middle East and North Africa. Another is African Telecommunications Limited (AFRICOM) of Atlanta for a satellite to serve Africa. These are also in GSO and may use two satellites to cover the intended region. To date, details on these systems have not been made available. The expected coverage areas for these systems are shown in Fig. 8.7.

Hughes is building the satellite for ASC (Agrani) using their HS601 (or HS702) bus, and will use one antenna 30 ft in diameter for transmitting, and one 50 ft in diameter for receiving. The antenna will produce 250 beams with each approximately 400 km in diameter.

The ACeS system, being built by Lockheed Martin, will use dual antennas roughly 12 m in diameter and operating at L-band. The design will provide 140 beams and will serve Southeast Asian countries and China. The footprints of the beams are indicated in Fig. 8.8.

The handsets will be dual mode for satellite-terrestrial cellular service. Single-hop connection between two user terminals is possible through the satellite. An artist's rendering of the Lockheed Martin spacecraft design is illustrated in Fig. 8.9. The system will be compatible with the GSM standard.

Figure 8.7
Regional geostationary mobile satellite systems providing service to handheld transceivers.

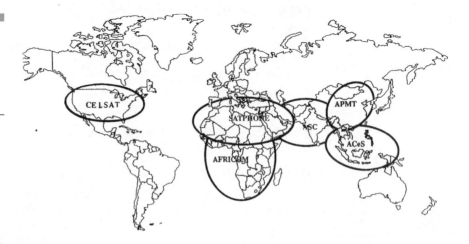

Figure 8.8
The 140 footprints laid down by the ACeS GSO regional satellite on Southeast Asia.

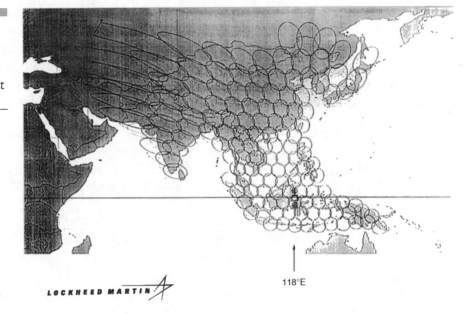

LOCKHEED MARTIN

118°E

● The footprints straddle the equator with most lying above the equator. The footprints are distended due to the curvature of the Earth. Several on the left partially fall off the Earth (below the horizon) at the limb. Earth terminal beam angles in this region equal zero degrees.

Figure 8.9
ACeS geostationary satellite configuration.

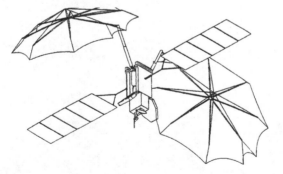

The large unfurlable antennas operate at L-band, and uplink at 1.6 GHz and downlink at 1.5 GHz.
Terrestrial antenna footprints(cells) are 400-500 meters in diameter.
Generates 140 terrestrial cells. Downlink power per cell is 20 watts.
The boom-mounted antennas are 12 meters in diameter, with the feeds mounted on the bus (body).
Dish shown in lower part of bus is a C-band antenna used to communicate with network control center and gateways (interface with PSTN).
Solar cells supply 10,000 watts of direct current power.
System uses on-board processing.

The antennas are unfurlable and placed at the end of booms. The feeds for the spot beams are located on the satellite bus (body) structure. The smaller unfurlable antenna will operate at 1.6 GHz and the larger one at 1.5 GHz. A single beam at C-band is used for up- and downlinking to gateways and the network control center.

The United States' regional CELSAT system will provide cellular service to the continental United States, Alaska, Hawaii, and Puerto Rico/Virgin Islands. The cellular coverage is shown in Fig. 8.10. This consists of 118 spot beams, and the footprints are 200 mi in diameter. Small cells allow efficient reuse of the spectrum, high channel density, and low transmitted power (the high EIRP results mainly from the high antenna gain).

8.4.2 Huge Antennas for GSO Cellular Operation

Because of the poor performance of the handheld transceivers, the satellite must manifest high performance to make up for this shortfall. That is, the space segment must transmit a high EIRP because of the small G/T (typically −25 dB/K) of the terrestrial receivers. In addition, because

Figure 8.10
Cellular structuring of the United States by spot beams from the CELSAT satellite in geostationary orbit. Service is to handhelds.

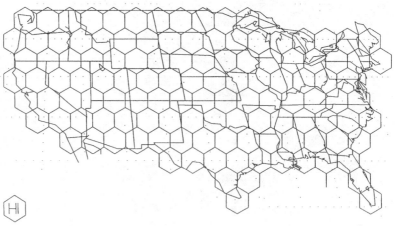

● 118 spot beams.

■■ ■ ■■ ■
Figure 8.11
Link equations
showing the require-
ments placed on an
MSS satellite in GSO
to provide service
to handheld
transceivers.

$P_r = P_t\,G_t\,G_{rec}\lambda^2/(4\pi R_{slt})^2 M$ Friis equation

$N = kT_0 B \bullet \overline{NF}$

$N_0 = kT_s$ (includes noise figure
and antenna noise temp.)

UP:

$$(C/N_0)_{up} = P_t\,G_t\,G_r\lambda^2/(4\pi R)^2 kT_s L$$

$$(C/N_0)_{dB} = \boxed{10\log P_t G_t} - 20\log(4\pi R/\lambda) + \boxed{10\log G_{Rs}/T_s} - 10\log k - 10\log L - M$$

Earth
station
EIRP

Satellite
figure of
merit

DOWN:

$$(C/N_0)_{dwn} = \boxed{10\log P_s G_s} - 20\log(4\pi R/\lambda) + \boxed{10\log G_r/T_r} - 10\log L - 10\log k - M$$

Satellite
EIRP

Earth
station
figure of
merit

Power-flux density (PFD) at sat. or earth station: $\phi = P_t\,G_t(\theta,\phi)/4\pi R^2$ W/m^2

Antenna gain: $4\pi A_p\eta/\lambda^2$, $A_{physical}\eta = A_{eff}$

$P_{received} = \phi \cdot A_{eff}(\theta,\phi) = \left[P_t\,G_t(\theta,\phi)/4\pi R^2\right]A_e(\theta,\phi)$

A_p: Physical aperture area (aperture antennas, not endfire or wire antennas)

N: Receiver noise power

N_0: kT_s noise power density

M: Margin power above the minimum requirements to ensure performance against any
propagation anomalies.

R_{slt}: Slant range from es/sat or sat/es.

L: Additional losses (e.g., polarization, pointing, and so forth).

of the low EIRP of the handset (<1 W watt peak), the satellite G/T must
be large. This condition can be satisfied by using a large spacecraft
antenna.

These shortfalls in the up- and downlinks are shown in Fig. 8.11. This
has been shown previously but is reproduced here for convenience. In the
uplink, the small EIRP of the earth terminal (in dotted box) is compensated by the large figure of merit (G/T_s) of the spacecraft. Similarly, on the
downlink, the high satellite EIRP (dotted box) makes up for the low figure of merit (G/T_r) of the earth terminal.

The large satellite antenna also lends itself to providing small beam
width multiple spot beams and ensuring larger power-flux density ϕ
incident on the earth as shown in the footnotes on Fig. 8.11.

8.5 Handheld Terminal Design Challenges

Up to this point we have stressed the space segment and the challenges of providing service to handheld terminals. There are problems enough in this area, but, in addition, the design of the handset transceivers also poses considerable problems. Some of the areas to be addressed in this design include:

- The handset must provide full duplex voice service with adequate signal fidelity, and in addition must be made to handle data, fax, paging, and possibly position determination.

- Physically small antenna provides for transmitting and receiving to/from the satellite. The attributes include coverage (beam width), gain, and polarization.

- Miniaturization of the RF and baseband components: this would include the use of monolithic microwave integrated circuits (MMIC) and very large scale integration (VLSI).

- Low-rate (and variable-rate) vocoders should provide near-toll quality voice, coupled with forward error correction (FEC) coding to enhance performance in a hostile environment.

- Dual-mode operation adaptable to both terrestrial and satellite-based cellular service allows subscribers to interconnect with terrestrial cellular networks. The supposition is that the standards used are compatible, for example, GSM, IS-54, IS-95, and AMPS. That is, global dual-mode operation imposes constraints due to difference in frequency of operation in the various ITU Regions (1, 2, and 3), access techniques, and modulation used in the overfly country.

- Batteries should offer adequate capacity, long standby/active life, and low weight.

- Establish limits on radiated power which is not injurious to human life.

For satellites in LEO and MEO, there are no discernible time delay problems which will cause double-talk (echoes) in the voice conversation. The raw propagation delays ($\tau = R/c$) are relatively small in comparison to delays for satellites in geostationary orbit. However, there are contributions to delay resulting from the circuits such as the vocoders. Vocoders may introduce delays (depending on bandwidth compression) up to about 80 ms.

For mobile applications, the antenna should provide coverage of the satellite no matter in what direction the transceiver is pointing. For handsets the only viable coverage is omnidirectional in azimuth near 180° in elevation. Therefore, since gain is inversely dependent on beam width, one must accept the low gain which may be in the range of 0 to 3 dB.

The LEO/MEO satellites will downlink using circular polarization. For the maximum received signal, the transceiver antenna should also be circularly polarized with the same sense circular polarization. Frequently the open literature shows a whip or stub antenna on the handsets. These are linearly polarized, and one can expect a 3-dB loss if illuminated by a circularly polarized signal. Circularly polarized antennas on both the handsets and the satellite should also have small axial ratios (small departure from circularity). This ensures high antenna efficiency and signal coupling.

Based on work funded by military programs, the RF componentry in the transceivers can be made physically small by using monolithic microwave integrated circuits (MMIC), and very large scale integration (VLSI) in the signal processing sections. Most of these circuits consume a small amount of power (save the power amplifiers) and both of these areas of technology are mature.

Vocoders built into handsets use low data rates. Vocoders which will compress conversations to rates of 4.8 kbps and 9.6 kbps are common. However, rates of 13 kbps are also being used, and all use some form of forward error correction and interleaving to cope with random and bursty errors, respectively. Pure data information may also be transmitted at lower rates.

Multimode operation is also being built into the handsets. The word multimode can be confusing. It may mean capable of operating with either the terrestrial cellular system or using the satellite route. Handsets are now made which will seek out the terrestrial cellular first, and if not available, revert to the satellite mode. Another definition of multimode is incorporating several digital or analog standards within the unit. This may include some form of TDMA or CDMA or the earlier analog AMPS system. Generally, terrestrial systems operating multimode refer to the latter definition.

Satellites which use bent-pipe transponders can accommodate the various access and modulation modes since they are wide open. Satellites which use onboard signal processing may be frozen in the type of signal format that can be used.

Batteries are still the highest weight component in the earth terminal. Even though advances have been made in the technology, much work is still to be done to improve the performance and reduce the weight.

The amount of power which can safely be radiated by a handset is still an argumentative issue. For handsets, peak power less than 1-W range is used. Portable and vehicle-mounted units can use more power since the transmitting antenna is not in proximity to the user's head, and the beam, being directional, is pointed away from the user.

Some performance numbers for the handhelds are as follows:

- *Antenna.* Omnidirectional coverage with adequate gain and using circular polarization. Gains are in the 0 to 3 dB range. Coverage area is quasi hemispherical.

- *Power.* Transmitted power below 500 mW average.* EIRP is in the range of −3 to 0 dBW.

- *G/T.* Receiver figure of merit is in the range −25 to −22 dB/K.

- *Battery life.* Greater than 4 h talk time, 24 h standby time is available.

- *Size.* Comparable to terrestrial size cellular phones.

Typical cellular handsets which will be used with two Big LEOs are shown in Fig. 8.12. The one on the left (Fig. 8.12*a*) will be used by Globalstar and will be using CDMA techniques. The uplink frequency will be at L-band, and downlink will be at S-band. The design is by Qualcomm. The antenna consists of two quadrifilar helices for L-band and S-band operation.

The unit on the right (Fig. 8.12*b*) is the Motorola design to be used with their Iridium system and is a dual operation handset for communications via terrestrial cellular or satellite. The unit will also be capable of being switched to several wireless infrastructure standards, depending on the country in which the user may be located. The module standard is inserted at the top of the unit.

For terrestrial cellular, a linearly polarized whip antenna (not shown) is telescopically deployed. For satellite operation, a quadrifilar helix antenna, using circular polarization, is folded up from the side of the unit. The antenna in the fold-down position is shown on the left side. The speaker is located on a fold-down flap shown at the bottom of the unit (not deployed in the figure).

The prices of the handsets may vary depending on the orbital regime with which they operate, single- or multimode operation, and the number of features. Prices from $100 (for Little LEO units) to $3000 and for those providing voice service have been mentioned.

---

* For nonhandheld terminals (portable, vehicle-mounted) the power may be in the 1 to 5 W range, with antenna gains around 10 dB, and *G/T,* is in the range of −15 to −20 dB/K.

Figure 8.12

Typical handset transceivers for LEO satellite-based cellular service.

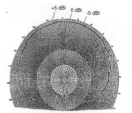

Magnitude dBi vs Elevation

Hand-Held Quadrifilar Helical
Antenna Radiation Pattern

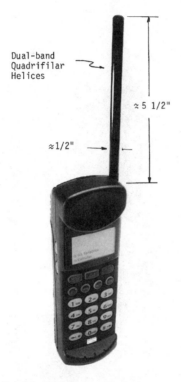

Dual-band
Quadrifilar
Helices

≈ 5 1/2"

≈1/2"

Not to scale.

a) Cellular Phone for
   Globalstar System
   Cost: ≈ $750

● Both are dual-mode transceivers.

b) Cellular Phone for
   Iridium System
   Market entry cost: ≈$3000
   Battery:1 hour talk time
           23 hour standby

Figure 8.13 represents two handsets which are being developed for satellites operating in GSO. They are being designed by Ericsson, Inc. The sketches were supplied to the author by Dr. Karabinis who works at the Ericsson, Inc. facility in Research Triangle Park, North Carolina. Both are dual-mode transceivers operating with terrestrial cellular and are satellite-based. When the units are first turned on, they will first seek terrestrial cellular communications, and if not available, will use the satellite mode (the former is preferred because the service may be

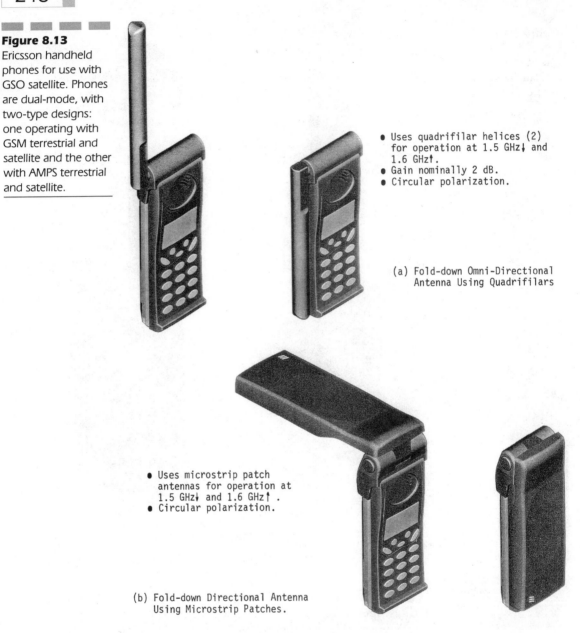

Figure 8.13
Ericsson handheld
phones for use with
GSO satellite. Phones
are dual-mode, with
two-type designs:
one operating with
GSM terrestrial and
satellite and the other
with AMPS terrestrial
and satellite.

- Uses quadrifilar helices (2)
  for operation at 1.5 GHz↓ and
  1.6 GHz↑.
- Gain nominally 2 dB.
- Circular polarization.

(a) Fold-down Omni-Directional
    Antenna Using Quadrifilars

- Uses microstrip patch
  antennas for operation at
  1.5 GHz↓ and 1.6 GHz↑ .
- Circular polarization.

(b) Fold-down Directional Antenna
    Using Microstrip Patches.

less expensive, and communications quality may be better). Two types
of handsets are being developed. One will operate with GSM terrestrial
and satellite. The other will operate with AMPS terrestrial and satellite.
Both are designed to operate at 1.5 GHz↓ and 1.6 GHz↑ in the satellite mode.

The unit at the upper left in Fig. 8.13a uses circularly polarized quadrifilar helices with omnidirectional/hemispherical coverage (see Chap. 7). The gain is nominally 2 dB. The antenna folds down against the case when not in use. The unit at the lower right (Fig. 8.13b) uses circularly polarized microstrip patches. One interesting aspect is that the main radiation is directed away from the user's head. Because of the beam's directionality, the unit (beam) may have to be slewed by body movement so that the beam faces the satellite.

References

Mitchell, H., "Mobile Telecommunications the Global Connection," Lockheed Martin, Vu-Graph presentation given at Villanova University, November 1996.

Rains, L., ed., *Space News,* "ACeS Backers Nearing Final Stages of Project," September 23—29, 1996.

Selding, P. B., "ESA Borrows Soviet Idea for New Satellite" (Archimedes), L. Rains, ed., *Space News,* May 1991.

Smith, B., "HS 601 to Link Hand Held Phones," D. M. North, ed., *Aviation Week & Space Technology,* January 30, 1996.

Various non-GSO Satellite applications submitted to the FCC.

Warren, R., ed., *Satellite Week* (Satellite Diary), "PT Asia Cellular Satellite (ACeS)," September 20, 1996.

GLOSSARY

ACeS   Asian Cellular Satellite.

ACI   Adjacent channel interference.

advancement of nodes   The eastward migration of the ascending node for an orbit in retrograde motion ($90° < i < 180°$). This movement subtracts from the longitudinal differential crossing of the satellite due to the earth's rotation.

AMPS   Analog mobile phone service; analog FM system, 30 kHz channels.

AMSC   American Mobile Satellite Consortium.

AMSC   American Mobile Satellite Corporation.

AMSS   Aeronautical mobile satellite service.

apogee   Point in the satellite elliptical orbit which is farthest from the center of the earth. The satellite assumes its lowest velocity in this part of the orbit (Kepler's second law).

argument   Archaic name for angle.

argument of perigee   The angle from the ascending node to perigee measured in the orbit plane. Note this differs from the longitude of perigee.

ascending node   The point along the orbit of the satellite where it crosses from south to north of the equatorial plane. Opposite side of the earth is the descending node.

ASEAN   Association of Southeast Asian Nations.

AWGN   Additive white gaussian noise.

balun   A network that converts an unbalanced line (coaxial cable) to a balanced antenna (wire antenna). The balun may assume different forms.

bent-pipe transponder   Satellite transponder which merely takes the uplink signal frequency and translates it to a lower frequency and then amplifies it for transmission. That is, there is no onboard processing.

BER   Bit error rate.

BFN   Beam-forming network.

BFSK   Binary frequency shift keying, no carrier recovery needed, no coherent receiver; two frequencies represent 0, 1; frequencies are orthogonal (separated by bit rate $1/T$); bandwidth efficiency <1 bps/Hz.

BPSK Binary phase shift keying; phase of transmitted carrier changes by 180° whenever the logic value of the binary data changes. When band-limited, the envelope goes to zero at the bit transitions and spectrum regrowth occurs after limiting or nonlinear application, and therefore linear amplifiers must be used to prevent regrowth.

CCI Cochannel interference.

CDMA Code division multiple access.

CPM Continuous phase modulation; basically a frequency modulation in which the baseband signal is a digital signal instead of a familiar signal as voice, music, or TV; phase is continuous at symbol boundaries; no jump in phase like in PSK; reduces spectral sidelobes and thus ACI and ISI; constant envelope; MSK, GMSK, and TFM in this family.

direct motion The term applied to eastward or counterclockwise motion of the satellite as seen from the north pole. Motion where $0 < i < 90°$. Sometimes called a posigrade orbit.

docket number Assigned to a proceeding if an NOI or NPRM is issued by the commission.

ecliptic plane Plane of the earth's orbit about sun.

EIRP Effective isotropic radiated power.

ETDMA Extended TDMA (6 users/30 kHz ch, uses vocoder: 4 kbps).

excess path loss Loss over and above free space loss due to multipath and shadow effects.

fade margin In the presence of fading the extra power required to achieve the desired bit error probabilities or performance.

FDMA Frequency division multiple access.

first point of Aries Another name for the vernal equinox. Point of Aries probably should now be called the point of Pisces, since the sun is now in the constellation Pisces at the spring equinox. Two thousand years ago, it was in Aries.

frequency reuse Reuse of the allocated frequency without causing interference. For example, use of the same frequency in spatially isolated spot beams or the use of opposite sense circular polarization in the two signals. Also, using the same frequency band several times in a coverage area of a cellular cluster, so as to increase the capacity of the system, and all within the allocated spectral bandwidth.

FSS Fixed satellite service.

gateways Provide the interface between low orbiting satellites and the public switched telephone networks (PSTNs).

GMSK Gaussian MSK; CPFSK with gaussian shaped pre-FM modulation filter; substantially suppresses out-of-band spectrum (function of LPF bandwidth), thus permitting reduced channel spacing; a modest cost in BER; used in European cellular GSM; constant envelope.

GSM Global system for mobile communications (Europe) first digital cellular system.

GSO Geostationary orbit.

handheld See handset.

handset, handheld, personal communicator All mean basically the same item. Pocket-size transceivers communicating full duplex with another user via the satellite. Employs antennas which usually provide omnidirectional coverage.

Hata model A terrestrial cellular propagation model in which the signal (none line of sight to receiver) to the receiver is attenuated by roughly a $1/R^4$ factor and not the free propagation model factor $1/R^2$.

HEO Highly elliptical orbit.

ICI Interchannel interference.

ICO Intermediate circular orbit.

IOC Intermediate orbit communications.

IS-54 U.S. standard, DAMPS, TDMA uses 3 chs in one AMPS channel (30 kHz).

IS-95 A U.S. standard, CDMA access with probable QPSK modulation (note CDMA is *not* the modulation); uses RAKE.

ITU International Telecommunications Union.

K factor Power in the direct line-of-sight component power plus specular component power divided by the fading Rayleigh component power. Frequently given in terms of 10 log [direct + specular/diffused]

Kraus helix A single winding helix producing circular polarization and fed against a ground plane.

LEO Low earth orbit.

LHCP Left-hand circular polarization.

linear modulation The amplitude of the transmitted signal $s(t)$ varies linearly with the modulating digital signal $m(t)$. These signals are bandwidth efficient but not constant envelope after band limiting. Therefore, they need to be linearly amplified (after filtering) to prevent spectrum regrowth. Included in this category are BPSK, QPSK, OQPSK, and π/4-QPSK.

lognormal distribution   The distribution of a positive variable whose logarithm has a gaussian distribution.

LMDS   Local multipoint distribution system.

LMSS   Land mobile-satellite service.

***M*-ary PSK**   Multipoint phase shift keying; circular constellation in phase space; constant envelope signal; for example, 2 psk-two antipodal points in signal phase space, not constant envelope.

***M*-ary QAM**   Multipoint quadrature amplitude modulation; usually rectangular constellation (need not be) in signal phase space; nonconstant envelope signal, penalized when transmitted through a nonlinear channel; good bandwidth efficiencies, but not power efficient.

MCHI   Mobile Communications Holdings, Inc.

mean motion ($2\pi/T$)   The value of the constant angular velocity for a spacecraft to complete a revolution.

MEO   Medium Earth Orbit.

MMIC   Monolithic Microwave Integrated Circuit.

MMSS   Marine mobile-satellite service.

MO&O (Memorandum opinion and order)   Issued by the commission to deny a petition for rule making, conclude an inquiry, modify a decision, or deny a petition for reconsideration of a decision.

MSK   Minimum shift keying; FM with modulation index 0.5; constant envelope; similar to staggered QPSK with one-half sinusoidal filter weighting; phase continuity in the RF carrier at the bit transitions; more spectrally efficient than BPSK, QPSK, and OQPSK; minimum separation between binary frequencies (separated by half the bit rate $\frac{1}{2}T = R/2$) representing the bits; does manifest out-of-band radiation; spectral rolloff $1/f^4$; Sometimes referred to as fast FSK (FFSK). Uses full response signalling [each pulse (bit) exists for only I (width of bit)].

MSS   Mobile satellite service.

nadir   Point in the satellite orbit which lies directly below the satellite.

Nakagami-*m* distribution   A generalized distribution that can model several different fading environments (Rayleigh, Rician, and lognormal) depending on the bounds placed on the parameters chosen.

Nakagami-Rician distribution   Represents the distribution of the length of a vector which is the sum of a dominant or fixed vector and a vector whose length has a Rayleigh distribution.

NF Noise figure.

NGSO Nongeostationary orbit.

NOI (Notice of inquiry) Issued by the commission (FCC) when it is simply asking for information on a broad subject or trying to generate ideas on a given topic.

non-GSO Nongeostationary Orbit satellites (LEO, MEO, HEO).

NPRM (Notice of proposed rule making) Issued by the commission when there is a specific change to the FCC rules being proposed.

oblateness Flattening of the earth caused by its speed of rotation.

OMT Ortho Mode Transducer.

OQPSK Offset QPSK; more robust against degradation in filtering and nonlinear processing than QPSK (no 180° phase changes; same spectral rolloff as QPSK; little spectral restoration after band limiting and limiting; sometimes referred to as staggered QPSK.

PCN Personal Communications Networks.

PCS Personal Communications Service.

perigee Point in the satellite elliptical orbit which is closest to the center of the earth. This is the perigee distance r_p. The perigee altitude is h_p.

personal communicator See handset.

π/4-QPSK Two quadrature PSK; offset by 45°, no ±180° transitions (good) like in PSK, but ±45°, ±135°; not constant envelope; used in American IS-54 and Japanese PDC system (cellular). Can use nonlinear amplifiers (Class-C) with negative feedback to reduce distortion.

polar orbit An orbit that passes over or nearly over the poles. The inclination angle is nominally 90°. Note that the nodal precession is zero for $i = 90°$.

precession Orbit migration eastward (retrograde orbit) or westward (direct orbit) resulting from the oblateness of the earth.

PSTN Public switch telephone network.

QPSK Quaternary PSK; not very popular in terrestrial and satellite communications; not constant envelope prior to bandlimiting, thus not suitable for use in nonlinear channels, if band-limited. Limiting restores constant envelope but also restores sidelobes; phase changes can be 0°, ±90°, or ±180°; spectral rolloff $1/f^2$ (6 dB per octave).

quadrature hybrid A device which produces two equal amplitude signals but differing in phase by 90°.

quadrifilar helix   A 4-winding helix with a 90° separation between elements. Produces circular polarization. May work with or without a ground plane.

RAAN   The right ascension of the ascending node. The angle measured eastward in the equatorial plane from the first point of Aries to the ascending node.

RAKE   Not an acronym, a network which "rakes in" multipath signals and improves the performance; used in IS-95.

RAKE receivers   An ensemble of correlators which intercepts the multipath signals and adds them and contributes to the bona fide (direct) output signal.

R&O (Report and order)   Issued by the commission to state a new or amended rule or to state that the FCC rules will not be changed.

RAS   Radio astronomy service.

Rayleigh fading   Multipath diffused fading which can be modeled by Rayleigh probability distribution.

RDSS   Radio determination satellite service. RF transmissions from the satellite(s) which enable terrestrial transceiver users to determine their location. This service may also be provided by the Global Positioning Satellite (GPS), for example, as that used in the ORBCOMM Microstar system.

regression of nodes   The westward migration of the ascending nodal point. Adds to the equatorial plane crossing due to the earth's rotation. The regression is manifested for direct orbits $(0 < i < 90°)$.

retrograde motion   Westward or clockwise motion of the orbit as seen from the north pole $(90° < i < 180°)$.

RF   Radio frequency.

RHCP   Right-hand circular polarization.

RM (Rule making)   Number assigned to a proceeding after Bureau/Office review of the petition for rule making, but prior to any commission action on the petition.

S&F   Store and forward.

SMSK   Serial MSK; constant envelope; less degradation by carrier phase error than SQPSK; serial processing rather than parallel processing; used in NASA's ACTS satellite.

sun-synchronous   A LEO orbit which rotates in synchronism with its motion about the sun in its yearly excursion. This is basically a nodal advancement which amounts to 0.986° per day or 360° per solar year.

sun-synchronous twilight zone orbit   A synchronous orbit which is always perpendicular to the earth-sun line, and will always remain in sunlight.

Suzuki fading   A mixture of the Rayleigh distribution and the lognormal distribution.

TCM   Trellis coded modulation; not a modulation scheme in itself but a combination of coding and modulation; achieves coding gain *without* the need for additional bandwidth; a breakthrough in communications.

TDMA   Time division multiple access.

TEC   Total electron columns.

TFM   Tamed frequency modulation, spectral sidelobes practically zero; better spectrum than MSK; $B_b T \approx 0.2$; constant envelope. Uses partial response signaling, and each filtered bit exists for several bit periods ($0 \leq t \leq LT, L > 1$).

true anomaly   Angle measure from the direction of perigee to the radius vector of the satellite in orbit, and measured in the direction of the satellite motion. Usually designated by the Greek letter ν (see Fig. 2.13).

two-body orbit   The unperturbed motion of a body of negligible mass (e.g., satellite) around a center of attraction (earth).

Van Allen belts   Two toroidal-shaped radiation belts of atomic particles encircling the earth. Detrimental to satellite's electronics. Inner belt: 1000 to 5000 km. Outer belt: 15,000 to 25,000 km.

VITA   Volunteers in technical assistance.

VLSI   Very large scale integration.

VSWR   Voltage standing wave ratio.

WARC   World Administrative Radio Conference.

WRC   World Radiocommunications Conference.

INDEX

ABOUT THE AUTHOR

Bruno Pattan is a Senior Member of Technical Staff with the Federal Communication Commission's (FCC) Office of Engineering and Technology in Washington, D.C. He has published over 70 articles and reports in the field, and has served as an instructor for many courses in satellite communications.